Fundamentals of Fruit Crops

The Authors

Dr Jagendra Pratap Singh got enrolled in BSc. (Ag.) at Govind Ballabh Pant University of Agriculture and Technology, Pantnagar (Uttarakhand) through ICAR and further completed MSc. (Ag.) in Horticulture from Narendra Dev University of Agriculture and Technology, Faizabad (Uttar Pradesh) and Ph.D. (Ag.) Horticulture form Chandra Shekhar Azad University of Agriculture & Technology, Kanpur (Uttar Pradesh). Dr Singh is Author of a Book and Published more than 7 Research Papers and 8 Articles, in Journals of National and International Seminars, Symposiums, Conferences. He also has received Best Thesis Award by Scientific society. He is currently working as FSO under Government of Uttar Pradesh.

Dr Saurabh Tomar completed his M.Sc. Agriculture and Ph.D. Agriculture in the year 2015 & 2018 respectively, from the Department of Horticulture, Chandra Shekhar Azad University of Agriculture and Technology, Kanpur (Uttar Pradesh). He secured First Rank in UPCATET Ph.D. Entrance Examination in 2015. Dr Tomar has authored a Book and published more than 15 Research Papers, 12 Articles in Journals of National and International Repute, Seminars, Symposiums, Conferences. Dr. Tomar is recipient of **"Young Scientist Award"**, **"Associate Young Scientist Award"**, **"Young Researcher Award"**, and Best Oral Presentation Award for his contribution in research and extension activities in the field of horticulture by various scientific societies. He is currently working in Chandra Shekhar Azad University of Agriculture and Technology, Kanpur (Uttar Pradesh).

Fundamentals of Fruit Crops

by

Jagendra Pratap Singh

Saurabh Tomar

2020

Associated Publishing Company®

A Division of

Astral International Pvt. Ltd.

New Delhi – 110 002

ISBN: 9789390371549 (Int Edition)

Publisher's Note:

Every possible effort has been made to ensure that the information contained in this book is accurate at the time of going to press, and the publisher and author cannot accept responsibility for any errors or omissions, however caused. No responsibility for loss or damage occasioned to any person acting, or refraining from action, as a result of the material in this publication can be accepted by the editor, the publisher or the author. The Publisher is not associated with any product or vendor mentioned in the book. The contents of this work are intended to further general scientific research, understanding and discussion only. Readers should consult with a specialist where appropriate.

Every effort has been made to trace the owners of copyright material used in this book, if any. The author and the publisher will be grateful for any omission brought to their notice for acknowledgement in the future editions of the book.

Published by : **Associated Publishing Company®**
A Division of
Astral International Pvt. Ltd.
– ISO 9001:2015 Certified Company –
4736/23, Ansari Road, Darya Ganj
New Delhi-110 002
Ph. 011-43549197, 23278134
E-mail: info@astralint.com
Website: www.astralint.com

Digitally Printed at : **Replika Press Pvt. Ltd.**

Dr. Basant Ram
Former Vice Chancellor

Rajendra Agriculture University
Pusa, Samastipur, Bihar
and
N. D. University of Agriculture and Technology, Kumarganj, Faizabad, Uttar Pradesh

FOREWORD

Agriculture has been foundation stone of human civilization, since time immemorial. There are over 7.4 billion people vying for food currently on our planet. Further, the world population is growing rapidly and more than 1, 00,000 people are added to this every day. Thus, the global population by 2050 may hit 9.5 billion mark. To meet food requirement of above population, the additional requirement may be 70 percent more. There is little scope to raise the total food production through vertical yield increase. The productivity of major food crops like rice, wheat, maize, soybean and pulses has reached to al plateau and increase in their productivity level has very little scope. In such circumstances, diversification in agriculture with laying greater emphasis on promotion of horticultural crops is prime need of the day. In initial years of plantation, the added advantage can be derived by growing suitable intercrops that may provide additional income to the growers.

The entitled **"Fundamental of Fruit Crops"** by **Dr. Jagendra Pratap Singh** and **Dr. Saurabh Tomar** is a timely publication. The book contains detailed description about cultivation of various fruit crops with focus on varieties, agro practices, pest management, *etc.* I am sure; the book would of great use to students, teachers, researchers and functionaries of allied developmental departments under Govt. sector.

My big compliments to both the authors for their splendid efforts and wish them all the best for their future endeavor..

(Basant Ram)

Preface

Fruit is backbone of the Horticulture. As the spectrum covered by these crops is too wide, the students had to study different book, assimilate their contents and prepare suitable notes. Often this task is not easy and in case of less time and shortage of books it is really impossible. Keeping these facts in mind we had made an effort to make available a book suitable for preparation and revision for competitive examinations, at the same time maintaining academic rigour and faculty accuracy. All the chapters in the book had been properly covered according to their families and relative importance. Questions given in the book are very relevant to the competitive examinations and most of the previous year question asked in different exams had been incorporated. This book fulfils the need of a good reference-cum text book for the student community.

We express our sincere thanks to scientists & seniors from CSA University of Agriculture & Technology, Kanpur for their inputs in terms of material collection, compilation and correction. Every effort has been made to ensure that the information collected from various sources such as individuals, institutions, organizations and publications are accurate. We hope that this book will be a useful reference material for the teachers, practitioners and students searching their careers in ICAR Institutes, Indian agriculture and horticulture universities, State Agriculture/Horticulture Department and for competitions like State Public Services/UPSE/ARS/JRF/SRF/NET along with entrance examination for B.Sc., M.Sc. and Ph.D. Degree in Horticulture.

We hope that this effort will help the readers in their endeavours. All the suggestions in making the book further informative are always welcome.

Jagendra Pratap Singh

Saurabh Tomar

Contents

Chapter 1
Mango

Botanical name : *Mangifera indica* L.
Family : Anacardiaceae
Origin : Indo-Burma
Type of fruit : Drupe
Edible Portion : Mesocarp
X = 10
2n = 40

SOIL, CLIMATE, PLANTING, HIGH DENSITY PLANTING, NUTRIENT AND WATER MANAGEMENT, INTERCROPPING, OFF-SEASON PRODUCTION

Mango *(Mangifera indica)*, the king of fruits, is grown in India for over 400 years. India shares about 56% of total mango production in the world. Its production has been increasing since independence, contributing 39.5% of the total fruit production of India. Andhra Pradesh tops in total production, whereas Uttar Pradesh tops area-wise. Andhra Pradesh, Uttar Pradesh, Bihar, Karnataka, Maharashtra, West Bengal and Gujarat together contribute for about 82% of the total production in India.

Climate and Soil

Mango can be grown on a wide variety of soils under varied climatic conditions. It can be grown from alluvial to lateritic soils except in black cotton soil having poor drainage. It grows well in soils with slightly acidic pH. It does not perform well in soils having pH beyond 7.5. Soils having good drainage are ideal for mango. Mango is a tropical fruit, but it can be grown up to 1,100m above mean sea level. There should not be high humidity, rain or frost during flowering. The temperature between 24 and 27°C is ideal for its cultivation. Higher temperature during fruit development and maturity gives better-quality fruits.

The areas experiencing frequent showers and high humidity are prone to many pests and diseases. Thus it can be grown best in regions with a rainfall between 25cm and 250cm. Regions having bright sunny days and moderate humidity during flowering are ideal for mango growing.

Varieties

India is the home of about 1,000 varieties. Most of them are the result of open pollination arisen as chance seedlings. However, only a few varieties are commercially cultivated throughout India.

Commercial Mango Varieties Grown in Different States

- ☆ Andhra Pradesh: Banganapalli, Suvarnarekha, Neelum and Totapuri
- ☆ Bihar: Bombay green, Chausa, Dashehari, Fazli, Gulabkhas, Kishen Bhog, Himsagar, Zardalu and Langra
- ☆ Gujarat: Kesar, Alphonso, Rajapuri, Jamadar, Totapuri, Neelum, Dashehari and Langra
- ☆ Haryana: Chausa, Dashehari, Langra and Fazli
- ☆ Himachal Pradesh: Chausa, Dashehari and Langra
- ☆ Karnataka: Alphonso, Totapuri, Banganapalli, Pairi, Neelum and Mulgoa
- ☆ Madhya Pradesh: Alphonso, Bombay Green, Dashehari, Fazli, Langra and Neelum
- ☆ Maharashtra: Alphonso, Kesar and Pairi
- ☆ Punjab: Chausa, Dashehari and Malda
- ☆ Rajasthan: Bombay Green, Chausa, Dashehari and Langra
- ☆ Tamil Nadu: Alphonso, Totapuri, Banganapalli and Neelum
- ☆ Uttar Pradesh: Bombay Green, Chausa, Dashehari and Langra
- ☆ West Bengal: Fazli, Gulabkhas, Himsagar, Kishenbhog, Langra and Bombay Green.

State-Wise Availability of Mango in India

- ☆ Andhra Pradesh: March to mid – August
- ☆ Bihar: May-end to mid-August
- ☆ Gujarat: April to July
- ☆ Haryana: June to August
- ☆ Himachal Pradesh: mid-June to mid- August
- ☆ Karnataka: May to July
- ☆ Madhya Pradesh: Mid-April to July
- ☆ Maharashtra: April to July
- ☆ Rajasthan: May to July
- ☆ Tamil Nadu: April to August
- ☆ Uttar Pradesh: Mid-May to August
- ☆ West Bengal: May to August

In India, mango is available from March to mid-August. The north Indian cultivars are alternate-bearer whereas south Indian ones are generally regular-bearer.

About 20 varieties are grown commercially.

Alphonso

One of the most popular variety of India, it is mainly grown in Ratnagiri area of Maharashtra and to a small extent in parts of south Gujarat and Karnataka.

Its fruits are medium-sized (250g), with attractive blush towards the basal end. Pulp is firm, fibreless with excellent orange colour. It has good sugar: acid blend. Keeping quality is good. It is susceptible to spongy tissue.

Banganapalli

A widely cultivated, early-maturing mango of south India. It is the main commercial variety of Andhra Pradesh. Its fruits are large-sized, weighing on an average 350-400g. The pulp is fibreless, firm and yellow with sweet taste. Fruits have good keeping quality.

Bombay Green

It is one of the earliest varieties of north India. Its fruits are medium-sized, weighing about 250g each. Fruits have strong and pleasant flavour. Pulp is soft and sweet.

Chausa

Late-maturing variety of north India, it matures during July or beginning of August. Fruits are large, weighing about 350g each. Fruits are bright yellow with soft'and sweet pulp. It is shy bearing.

Dashehari

One of the most popular variety of north India, it is a mid-season mango. Fruits are medium-sized, with pleasant flavour, sweet, firm, and fibreless pulp. Stone is thin and keeping quality good.

Fazli

This is indigenous to Bihar and West Bengal. Fazli is a late-maturing (August) mango. Fruits are large, with firm to soft flesh. Flavour is pleasant and pulp is sweet and fibre less. Keeping quality is good.

Gulab Khas

It is indigenous to Bihar. Regular and heavy-bearer, it is mid-season mango. Fruits are small to medium-sized. It has rosy flavour. Fruits are ambre-yellow with reddish blush towards the base and on sides. Keeping quality is good.

Himsagar

Very popular in West Bengal, it is a regular-bearing mango. Its fruits are medium-sized, having good quality. Flesh is firm, yellow, fibreless with pleasant flavour. Keeping quality is good.

Kesar

Popular in Saurashtra region of Gujarat, Kesar is an irregular-bearing mango. Fruits are medium-sized. Flesh is sweet and fibreless. It has excellent sugar: acid blend. Fruits ripen to attractive apricot-yellow colour with red blush. It has good processing quality.

Kishenbhog

Indigenous to West Bengal, it is a mid-season mango. Fruits are medium to large-sized, good with a pleasant flavour. There are traces of turpentine. Flesh is firm with few fibres. Keeping quality is good.

Langra

An important commercial mango variety of north India, it is biennial-bearer and a midseason variety, with good quality fruits. Flesh is firm, lemon-yellow in colour and scarcely fibrous. It has characteristic turpentine flavour. Keeping quality is medium.

Mankurad

It is a mid-season variety, popular in Goa. Fruits are medium-sized with yellow skin. Flesh is firm, cadmium yellow and fibreless. Keeping quality is good.

Neelum

A heavy-yielding, late-season mango in south India, it has regular-bearing habit. Fruits are medium-sized with good flavour. Flesh is soft, yellow and fibreless. Keeping quality is good.

Pairi

A native to coastal Maharashtra including Goa, it is an early-maturing, heavy and regular-bearer mango. Fruits are medium-sized with good quality. It has good flavour with sugar: acid blend. Flesh is soft, primuline-yellow and fibreless. Keeping quality is poor.

Totapuri

Widely grown in south India, Totapuri is a regular and heavy-bearing mango. Fruits are medium to large with prominent sinus. Fruit quality is medium. It has a typical flavour and flat taste. Flesh is cadmium-yellow and fibreless.

A number of selections/hybrids of mango have been evolved. These include Clone C-51 from Dashehari selected at the CISH, Lucknow, and an off-season selection, Niranjan, selected at Parbhani. New clonal selections from Langra and Sunderja have been made at Varanasi and Rewa. A clonal selection, Paiyur 1, has been made from Neelum, in addition to few dwarf polyembryonic selections made in the north-eastern region. As a result of systematic hybridization, several hybrids have been released. However only a few have become commercially acceptable. Of these, Mallika, Ratna and Arka Puneet are becoming quite popular.

Propagation

Mango is a highly heterozygous and cross-pollinated crop. There are 2 types of mango varieties. Most of the varieties in south are polyembryonic and thus give true-to-type seedlings. In north, the varieties grown are monoembryonic and need to be propagated vegetatively.

Mango is propagated on mango rootstock. For raising rootstock, the seeds of mango are sown within 4-5 weeks after extraction otherwise they lose their viability. For sowing the seeds, raised beds are prepared with a mixture farmyard manure, red soil and sand. In some places, seeds are sown directly in polythene bags. After germination, the leaves turn green in 2-4 weeks.

These seedlings are transplanted to polythene covers containing red soil, sand and farmyard manure. Addition of nitrogenous fertilizer to polythene covers after the establishment of plants helps in quick growth of seedlings. The seedlings thus raised should be used for grafting at different ages. Several methods of grafting are practiced. They are:

- ✰ **Inarching:** It is one of the most widely practiced methods of grafting. One can get a big-sized plant material for planting with over 95% success rate.
- ✰ **Veneer and side grafting:** These can be utilized for preparing a grafted plant material or for *in situ* grafting, *i.e.* for the rootstocks which are already planted.
- ✰ **Epicotyl/stone grafting:** This method is widely practiced in the Konkan region of Maharashtra.

The germinated seedlings of 8-15 days old are used for grafting.

Cultivation

Planting

Different systems of planting like square, rectangular and hexagonal are followed at different places. However, square and rectangular systems are also popular. The spacing depends on the vigour of the variety and the cropping system. The planting season varies from Jun to September.

The main field is brought to fine tilth. Pits of 1m x 1mx 1m size are dug. These are exposed to sun for about 30 days. Before planting, pits are filled with well-rotten farmyard manure. The top and sub-soil are taken out separately while digging the pits. The grafts should be planted during rainy season. In the *in-situ* grafting, rootstocks are planted in the main field. Then they are raised for 6 months to 1 year. Then the scions of the variety that need to be grown are taken and grafted. This is usually done when humidity is high. After grafting the scions are covered with polythene covers.

High-density Planting

High-density planting helps increase the yield/unit area. In north India, mango Amrapali is found amenable for high-density planting with a spacing of 2.5m x 2.5m. Soil drenching with paclobutrazol (2 ml/tree) induces flowering during off year. It has become a commercial practice in Konkan region of Maharashtra. If coupled with pruning, it, helps increase production/unit area in Dashehari. The polyembryonic mango Vellaikolumban when used as rootstock imparts dwarfing in Alphonso.

After Care and Management

Training and Pruning

Training is an important practice during the first few years after planting. It is essential to space the branches properly to facilitate intercultural operations.

Manuring and Fertilization

The nutritional requirement of mango varies with the region, soil type and age. A dose of 73g N, 18g P_2O_5 and 68g K_2O/year of age from first to tenth year and thereafter a dose of 730g N, 180g P_2O_5 and 680g K_2O should be applied in 2 split doses during June-July and September-October, respectively.

Spraying of zinc sulphate (0.3%) during February, March and May is recommended to correct the zinc deficiency. Spraying of Borax (0.5%) after fruit set twice at monthly intervals and 0.5% manganese sulphate after blooming corrects boron and manganese deficiencies, respectively.

Organic manures and phosphatic fertilizers should be applied immediately after harvest, whereas ammonium sulphate should be given before flowering.

Intercropping

In mango, intercropping helps check weed growth and reduces nutrient losses. Intercropping blackgram-wheat-mango and brinjal-onion-mango gives better monetary benefits. Besides, taking up cover crops like sunhemp, cowpea, pea help to prevent soil erosion.

Irrigation

The young plants upto 2-year-old should be watered regularly. The newly-planted grafts need about 30 litres of water every week. Irrigation during pre-flowering phase increases flowering. Irrigating grown-up trees after fruit set at 10-day interval increases the yield.

Harvesting and Post-harvest Management

Mangoes should be harvested with pedicel. Injury to the fruits during harvesting brings down their quality and also makes them prone to fungal attack. An average mango tree yields 8 tonnes/ha. The number of fruits per tree doing its bearing age generally various from 1000 to 2000 fruits. The productivity of mango is higher in Andhra Pradesh and Bihar. The north Indian mangoes Langra and Dashehari are alternate-bearers, whereas most of the south Indian mangoes are regular bearers. Mango Mallika and Amrapali are also comparatively regular-bearer.

After harvesting, mangoes are graded according to their size. To maintain the quality, proper packaging is a must. In western region, bamboo baskets are used for packing. A basket contains 50-100 fruits. Straw is used for packing. Wooden boxes are also used in some place.

However, now perforated cardboard are generally used. In these boxes either fruits are individually wrapped with tissue paper before packing or paper shavings are used for cushioning. Minimizing the post harvest losses is one of the most important aspects. Usually green and mature mangoes are stored better than ripe ones harvested from trees. Low temperature storage, controlled atmospheric storage, use of chemical treatment for delaying ripening, irradiation, heat treatment, packaging and shrink wrapping are methods to increase their shelf-life.

The temperature of 5-16°C for different varieties is ideal for storing. Mangoes are highly susceptible to low temperature injury. Loss of flavour and development of undesirable softening are major symptoms of chilling injury. Under controlled atmospheric storage, retardation of respiratory activity, delaying of softening, colour development and senescence of fruits take place. Hence, this method has not been adopted in mango. The combination of waxing (3%) along with hot-water treatment results in good quality fruits with extended storage life. Individual wrapping of fruit imparts uniform colour and reduces shrinkage. Hydro-cooling at 12°-15°C and holding for 2 weeks at 15°C followed by storage for 1 week at ambient temperature gives good storage life to fruits.

Disorders

Alternate Bearing

Alternate bearing has been one of the major problems. Most of the south Indian varieties are regular-bearer, whereas north Indian ones alternate-bearer. Paclobutrazol is a promising chemical for flower induction in mango. Soil drenching with paclobutrazol (5g -10g/tree) results in minimum outbreak of vegetative flushes during September to October giving an early and profuse flowering and more annual yield without affecting fruit size and quality.

Mango Malformation

It is one of the most important disorders, causing huge losses. It is a major problem in Punjab, Delhi and Uttar Pradesh. However, it has also been noticed in Gujarat, Maharashtra, Bihar, West Bengal and Orissa. Of the 2 types of mango malformation, vegetative malformation is more common in nursery seedlings and young plants. Floral malformation affects trees at the bearing stage. In **vegetative malformation** or bunchy top, compact leaves are formed in a bunchbat the apex of shoot or in the leaf axil and growth of shootlet is arrested. **Floral malformation** directly affects the productivity. The incidence of disorder varies from variety to variety.

Deblossoming alone or coupled with a spray of 200ppm NAA lowers the number of malformed panicles significantly.

Black Tip

This disorder is mainly noticed in Punjab, Uttar Pradesh, Bihar and West Bengal. The distalend of the affected fruits turns black and becomes hard. These fruits ripen prematurely and become unmarketable. This disorder is caused by the smoke of brick-kilns located within a distance of 600m. Gases like carbon monoxide and carbondioxide, sulphur dioxide and acetylene cause these symptoms. It can be controlled by raising the height of the chimney of the brickkilns. Spraying borax (0.6%) at 10-14 days intervals starting from fruit set also controls it.

Clustering *(Jhumka)*

This malady is characterized by a cluster of fruitlets at the tip of the panicle giving an appearance of bunch tip called *jhumka*. These fruitlets are dark green with a deeper curve in the sinus beak region compared with normally developing fruitlets. These fruitlets grow to marble size after which their growth ceases. One of the main reasons for clustering is the adverse climate during February-March, particularly the low temperature. Most of the fruits are aborted with shrivelled embryos and do not develop further, signifying the role of normal embryo growth in the development of fruits.

Spongy Tissue

It is specific in Alphonso mango. Fruits from outside look normal. but inside a patch of flesh becomes spongy, yellowish and sour. This disorder has brought down the export of this variety. Inactivation of ripening enzyme due to high temperature, convective heat and post harvest exposure to sunlight are the causes. Use of sod culture and mulching are useful in reducing its incidence. Mango hybrids Ratna and Arka Puneet which have Alphonso like characters do not suffer from this malady. Harvesting mangoes when they are three-fourths matured rather than fully matured ones also reduces this malady.

Use of Growth Regulator

Due to the various causes, fruit drop occurs in mango rather at a higher rate, even upto about 99 percent in various stages of growth, more during the initial four weeks.

The extent of fruits drop can be reduced significantly by (a) Regular irrigation during the fruit development period (b) Timely and effective control measures against major pests and diseases, and (c) Through the application of growth regulators like NAA (50 ppm) and 2,4-D (20 ppm) during off years about six weeks after fruit set.

Pest and Diseases

- ☆ **Mango hopper (*Amaritodus atkinsoni*)** Spray phosalone @ 0.05% or carbaryl 2 g/lit or phosphamidon 1 m1/litre of water.

Nut weevil (*Cryptorrhynchus mangiferae* and *C.gravis*)

- ☆ General cleanliness in the orchard, destroying the adults in the bark crevices and holes and spraying with Fenthion 0.1%

Stem borer (*Batocera rufomaculata*)

- ☆ Padding with monocrotophos 36 WSC 10ml in 2.5 cm per tree soaked in absorbant cotton.
- ☆ Application of carbofuron 3G 5g per bore hole and plugging with mud.

Fruit fly (*Dacus* spp)

- ✰ Plough the inter spaces to expose pupae.
- ✰ Monitor with Methyl Eugenol traps.
- ✰ Remove the fallen fruits now and then and bury them deep into soil.

Powdery mildew (*Oidium mangiferae*)

- ✰ Apply sulphur dust (350 mesh) in the early morning to protect new flush or spray wettable sulphur 0.2% or Tridemorph 0.05%

Anthracnose and stalk and end rot (*Collectotrichum gloesporioides*)

- ✰ Spray Mancozeb 0.2% (1kg/ha) or Carbendezim 0.1% as preharvest spray, 3 times at 15 days interval.

Sooty mould (*Capnodium* sp.)

- ✰ Spray Dimecron 0.03% + Maida 5% (1kg maida or starch boiled with one litre of water and diluted to 20 litre).

Chapter 2
Banana

Botanical name : *Musa acuminata x* M.
Family : Musacea
Origin : South East Asia
Type of fruit : Berry
Edible Portion : Mesocarp and endocarp
X = 11
2n = 22, 33, 44

SOIL, CLIMATE, PLANTING, HIGH DENSITY PLANTING, NUTRIENT AND WATER MANAGEMENT, INTER CULTURAL SPECIAL OPERATIONS

- Banana is one of the oldest fruit known to mankind and also important food for man.
- **Origin:** South East Asia
- **'Apple of paradise'**
- Rich source of energy (137 K. Ca/100g)
- It is a good laxative.
- **Important states:** Tamil Nadu, Kerala, Maharashtra, Andhra Pradesh and Bihar.
- Edible bananas are mostly hybrids of the two species.
- *M. acuminata, M. balbisiana.*
- They set fruits by parthenocarpy.
- **Climate:** Humid tropic plant. Temperature range of 10°C to 40°C with an average of 23°C.
- **Altitude:** Upto 1500 mts from MSL.
- Wind velocity more than 80 m/hr will damage the crop heavily.
- **Rainfall :** 100 mm/month is good.
- **Soil:** Deep well – drained soil with abundant organic matter.

- ✰ Depth – one meter
- ✰ **Soil pH:** 5.5 – 8.0 found to be optimum.

Season of Planting

- ✰ Wet land – Feb-April: Poovan, Rasthali, Monthan
- ✰ April – May : Nendran, Robusta
- ✰ Garden lands : January – February and November – December
- ✰ Padugai lands : January – February and August – September
- ✰ Hill banana : April – May (lower palani hills)
- ✰ June – Aug (Sirumalai)

Propagation : Sucker

i. Sword sucker – suckers with a well – developed base and pointed tip having narrow sword shaped leaf bladers in the early stage.

ii. Water sucker or broad leaved sucker – small, undersized suckers of superficial origin bearing broad leaves.

- ✰ Sword suckers – more vigorous, grows faster and comes to bearing early.
- ✰ Average weight of the sucker – 1.5 to 2 kg.
- ✰ Micropropagation through tissue culture – Rapid multiplication of banana suckers.
- ✰ Pretreatment of sucker : The roots and decayed portion of the corn are trimmed.
- ✰ Pseudostem is cut leaving 20 cm from the corn.
- ✰ To avoid wilt disease infected portion of the corn may be pared, dipped for 5 min in carbendazion 0.1% (1 gm in 1 lit. of water) for wilt susceptible varieties – Monthan, Neyvannan, Virupahshi *etc.*
- ✰ **Pralinage** – with 40 g of carbofuran 3 G granules per sucker.
- ✰ (The corn is dipped in slurry solution of 4 parts of clay plus 5 parts water and sprinkled with carbofuran to control nematodes).
- ✰ Alternatively, dip the corn in 0.75% monocrotophas shade dried for atleast 24 hours and plant. Sow sunhemp on 45th day incorporate it after about a month. This operation reducem nematode build up.
- ✰ TC banana – plants with 5-6 leaves planting – *Pseudomonas fluorescence*/ plant 25 gm.

Field Preparation

The land is ploughed deeply and leveled. The pits of size 45 cm^3 is dug. The pits are refilled with top soil, mixed with 10 kg of FYM, 250 g of neem cake and 50 g of lindane 1.3%.

Spacing and Plants/ha

Garden land 1.8 x 1.8 m, 3086

1.5 x 1.5 m, 4444

Wet land 2.1 x 2.1 m, 2267

Hill 3.6 x 3.6 m, 750

- ☆ High density planting – 3 suckers/pit at a spacing of 1.8 x 3.6 m (4600 plants/ha).

Irrigation

- ☆ Irrigated immediately after planting, life irrigation – 4th day subsequent irrigation once in a week for garden land, 10-15 days in wetland after manuring.
- ☆ Drip irrigation – 15 lit/plant/day from planting to 4th month. 20 lit/plant/day from 5th to shooting and 25 lit/plant/day from shooting till 15 days prior to harvest.

Hill Banana

- ☆ 375 g of 40:30:40 NPK mixture and 130 g MOP/clump per application during October, January and April. Azospirillum and Phosphobacteria – 20 g each at planting and 5th month after planting preceding chemical fertilizer application.
- ☆ Apply N as neem coated urea. N & K in 3 splits 3rd, 5th and 7th month P at 3rd month of planting. For tissue culture banana 50% extra fertilizer at 2nd, 4th and 6th and 8th month after planting.
- ☆ For maximing productivity – fertigation.
- ☆ 25 litres of water/day + 200:30:300 g N:P_2O_5:K_2O/plant using water soluble fertilizer. For economizing the cost of fertilizers fertigate using normal fertilizers (urea and MOP) with 30% of the recommended dose along with recommended dose of P as basal at 2nd month of planting.

Interculture

- ☆ Mammutti digging at bi-monthly interval and earth-up
- ☆ De sucker – prune the side suckers at monthly interval
- ☆ Dry and dead, leaves are removed and burnt.
- ☆ Male flower – removed a week after opening of last hand
- ☆ Bunch emergence – propping. The trees are supported with bamboos or casurina poles to avoid damage by wind.

Growth Regulator

- ☆ Grade of bunch -2,4-D at 25 ppm (25 mg/lit) may be sprayed after the last hand has opened. This also helps to remove the seediness in poovan variety.

Spray CCC 1000 ppm of 4th and 6th month after planting. Spray plantozyme @ 2ml/lit at 6th and 8th month after planting to get higher yield.

Micronutrient

- $ZnSO_4$ (0.5%), $FeSO_4$ (0.2%), $CuSO_4$ (0.2%) and H_3BO_3 (0.1%) at 3, 5 and 7 MAP to increase yield and quality of banana.

Bunch Cover

- Use transparent polyethylene sleeves with 2% (during cool season)-4% (during summer season) ventilation to cover the bunches immediately after opening of the last hand.

Intercropping

- Leguminous vegetables, beet root, elephant foot yam and sunhemp. Avoid growing cucurbitaceous vegetables.

Physiological Disorders

Kottavazhai

- In certain pockets of Tamil Nadu, the banana cv. Poovan is manifested with a peculiar development disorder which is characterized by the presence of distinctly conical and ill filled fruits with a prominent central core having many under developed non viable seedy structures rendering the fruits inedible. This disorder can be overcome by spraying 2,4 D 20 ppm when the last hand of bunch is opened. The same chemical at same dose and same stage results in increased bunch weight and uniform grade especially in cvs. Nendran and Monthan.

Hard lump

- It is characterized by pinkish brown, firm pulp than the usual soft pulp occurs in cv.Rasthali, tastes like immature or unripe fruits. Spraying the bunches uniformly with 2,4 D at 1000 ppm or dipping the cut end of peduncle of the bunches for a period of 5 minutes appears to favour the reduction of lumps and improve the size.

Sunscald

- The peduncle of the bunches may be covered with flag leaf to prevent 'main stalk rot' and also the bunches with banana leaves to avoid sunscald.

Nematode

- Bunchy top virus – *Pentalonia nigronervosa.*
- Crop duration : bunches will be ready for harvest after 12-15 months of planting.

Viral Diseases of Banana

1. **Bunchy top** – transmitted by Aphid *Pentalonia nigronervosa*
   - Infected plants show short and narrow leaves together at the top of the pseudostem to form a bunch, hence this disease is known as 'Bunchy top'. The margins of leaves become wavy in advance stage of infection and roll upward.
   - **Management** : a) Remove all the affected plants along with complete rhizome, planting of virus free suckers.
   - b). control of banana aphid - spray 0.3% Rogar or Phosphomidon or Monocrotophos – 0.05% spray
2. **Banana bract mosaic virus** – transmitted by *Aphis gossypii*
   - *Pentalonia nigronervosa*
   - The name is derived from the conspicuous discolouration and necrotic streaks that develops on the bracts of the male bud. Spindle shaped discolouration found on the pseudostem.
   - **Management :** Effective control is similar to that of other viral diseases. It requires early detection, and immediate eradication of infected plants.
3. **Banana Streak -** Transmitted by citrus mealy bug –*Planococcus citri* Foliar symptoms resemble those of banana mosaic, especially in the early stages. Later, development of necrotic streaks

Fungal Diseases of Banana

1. **Panama wilt** *(Fusarium oxysporum f.* sp. *cubense)*
   - Yellowing of leaf blades, leaves wither and petiole breaks or buckles and hang around the pseudostem. Longitudinal splitting of pseudostem and subsequent death of entire plant.
   - **Management :** Removal of infected plants and application of lime @ 2kg/pit and leave it fallow for 6 months Growing resistant varieties like Dwarf Cavendish, Poovan and Nendran Crop rotation with paddy in wet land cultivation Capsule application of 50-60mg of carbendazim applied to the hole made at an angle of 45 degree diagonally in the diseases corm.
2. **Sigatoka leaf spot disease – *Mycosphaerella musicola***
   - Yellowish green streaks appear along the veins which later on enlarge into elongated/cylindrical spots. Several spots join together and cause drying of the leaves.
   - **Management:** Spray copper oxy chloride or carbendazim @ 500g/ha
   - Avoid close planting

Bacterial Diseases

1. **Moko wilt (*Pseudomonas solanacearum*)**
   - On leaves, yellowing starts from the inner leaf close to petiole and slowly spreads up ward. All the leaves turn yellow and wilting occurs.
   - **Management:** Good drainage facilities
   - Supress the wilt by bacteriazation with *P. inflorescens*
   - Crop rotation with sorghum
2. **Tip over or heart rot *(Erwinia carotovora)***
   - Seen mostly in tissue cultured plants . The middle tender leaf show rotting. Pseudostem easily comes out from corm portion
   - Bacterial oozing from edge of corm and pseudostem is also noticed
   - **Management:** Disease free suckers
   - Resistant variety Poovan can be grown

Harvest

- ✰ Bunches attain maturity from 100-150 days after flowering.

Yield (t/ha/year)

- ✰ Poovan – 40-50
- ✰ Monthan – 30-40
- ✰ Robusta – 50-60
- ✰ Dwarf Cavendish -50-60

Chapter 3
Citrus

CLASSIFICATION, SOIL, CLIMATE, VARIETIES, PLANTING, NUTRIENT AND WATER MANAGEMENT, INTERCULTURAL OPERATIONS

CITRUS

***Citrus* spp. Family : Rutaceae**

Citrus fruits include oranges lamons, limes, pummelo and grape fruit. Being a native of tropical and subtropical region of South East Asia, these have been under cultivation from time immemorial in South China, Malaya and sub-Himalayan parts of Assam, From here, they spread to other tropical and subtropical parts of the world. Next to mango and banana, citrus represents the third most important group of fruits in India.

The botanical classification of the genus is highly confusing since more and more inter specific and inter generic hybrids are going on added to the list each on deserving a separate species status.

All the edible fruits of citrus come under subgenus *Eucitrus* which can be divided into 5 horticultural groups.

1. Acid Group

Acid lime : Citrus aurantifolia

Tahiti or Persean lime : Citrus latifolia

Rangpur lime : C. limonia

Lemon : Citrus limon

Rough lemon : C. jambhiri

Citron : C. medica (Kidarankai in

Tamil, used for pickling)

Sweet lime : Citrus limettoides

2. Orange group

Sweet orange : Citrus sinensis

Sour orange : Citrus aurantium

(Narthankaai in Tamil, used for pickling)

Multiple leaf orange : C. multifolia

Japanese summer grape fruit : C. natsudaidai

3. Mandarin group : (loose jacket)

Coorg mandarin, Nagpur

Santra and Kodai orange : C. reticulata

Japanese Satsuma mandarin : C. unshiu

Willow leaf mandarin : C.deliciosa

King mandarin : C. nobilis

Kinnow mandarin : King x willow leaf

Tangerine orange var Dancy (trifoliate x mandarins) : Citrus tangerine

4. Pummelo and grape fruit group

Pummelo : C. grandis

Grape fruit : C. paradisi

Kumquat : Fortunella sp.

5. The fifth group consists of mainly hybrids of different citrus fruits with trifoliate orange (Poncirus trifoliata) and mainly used as rootstock.

e.g. Citrange (Poncirus trifoliata x C. sinensis) var. Troyer, var. Carrizo

Citrangor (Citrange x C. sinensis)

Tangelo (Tangerine x grape fruit)

Citrangequat (Citrange x kumquat)

Mandarinorange: *C. reticulata*

The group of orange is otherwise called Kamala orange. Nagpur santra of Maharashtra,bCoorg of Karnataka and Kodai orange of Tamil Nadu. This group is characterized by the loose skin of fruits.

Soil and Climate

- Subtropical 500-1500 m MSL elevation. A rainfall of about 150 cm to 250 cm is required. The winter should be mild and there should be no strong or hot wind during summer.
- A medium or light loam with a pH 5.5 to 6.5 would be ideal to grow.
- **Season :** November – December
- **Planting:** Seedlings and budded plants

- **Spacing:** 6 x 6 m pit size 75 cm^3 planting during May-June and September – October.
- Though the crop is grown as rainfed one, the young plants should be irrigated whenever there is failure of monsoon as well as during summer season.

Manure and Fertilizers

- Applied twice in a year during June and October.
- For Palani hills

Micronutrient

$ZnSO_4$ – 600 g

$MnSO_4$ – 600 g

$MgSO_4$ – 600 g

$FeSO_4$– 600 g

After Cultivation

- Removal of water shoots
- Rootstock sprouts
- Dead and diseased shoots
- Removal of laterals of the main stem upto 45 m from ground level
- Basins should be provided for each tree with gradient slope.

Growth Regulators

- To increase the fruit retention spraying the trees at flowering and again at marble stage with 2,4-D at 20 ppm or NAA 30 ppm.
- **Harvest:** Starts bearing from 3-5 year after planting in budded plants. Incase of seedlings 5-7 years.
- **Yield:** 15-20 t/ha/yr.
- A small crop can be obtained from 4 year old tree and the yield will be higher from 7th year. From flowering to maturity it takes 9 months.

Varieties

1. Kodai Orange (*Citrus reticulate*)

Trees are vigorous, fruits are very small characterized by loose rind and medium flavour. But it is a heavy seeded variety.

2. Nagpur Santra (*Citrus reticulata*)

It is the most important commercial cultivar of India. This cultivar is considered to be one of the finest mandarins in the world orange vigorous growing variety.

Fruits are medium to subglobose having a loose rind. Flesh is fine textured with abundant juice. Fruits mature in January – February.

3. Coorg (*Citrus reticulata*)

Medium to large fruits, bright orange with a loose rind. It has a good flavour, ripens later than Nagpur Santra. It is a regular bearer. It is the commercial variety in the coorg region of Karnataka.

In Darjeeling district of West Bengal, the variety grown is known as Darjeeling orange and it is Desi in Punjab. In Sikkim, the mandarin cultivar grown is known as 'Sumithira' while in Meghalaya it is called as 'Khasi Mandarin' or 'Sohniamtra'.

4. Sastuma Mandarin (*Citrus unshiu*)

It is a Japanese variety with small spreading tree. Fruits are seedless with thin rind having orange colour at maturity. The quality of fruit is excellent with good blend of sugar and acidity.

5. King Mandarin (*Citrus nobilis*)

It is a commercial variety of USA. The trees grow 5-6 M, petioles narrowly winged small, flattened, orange, red fruits with distinctly sweet pulp which is juicy and fexcellent quality.

6. Willow Leaf Mandarin (*Citrus deliciosa*)

It is another commercial variety of USA. Trees medium sized with drooping growth habit. The distinctive characteristics are presence of mild and pleasant aromatic flavour in fruit juice, plump spherical seeds, high degree of seed polyembryony and marked alternate bearing tendency of the trees.

7. Kinnow (King x Willow leaf)

It is a hybrid between king and willow leaf mandarins. It was developed by Dr. H.B. Frost at citrus Experiment Station, California 1915. It has performed very well in Pakistan, Punjab, Uthrangal, Haryana, Karnataka and foot hills of Himachal Pradesh. This hybrid cultivar produces excellent quality fruits and holds export potential. Fruit medium in size, globose to slightly oblat, rind thin rather adherent for a mandarin but peelable, tough and leathery, surface very smooth and glossy, colour yellowish orange at maturity.

Segments (9 to 10) do not separate easily, very juicy, flavour rich aromatic and distinctive, TSS 10o brin, acidity 0.8%. Seeds numerous, (2 to 2H) polyembryonic and cotyledonepale greenish yellow.

8. Dancy Tangerine (*Citrus tangerine*)

This is the commercial variety of Florida in united states. [(deliciola as male parent) experiment station California in 1915. It was first quality introduce in Punjab. Plants are medium to large, erect, symmetrical, dense foliage with a few scatted spaes, leaves broadly lanceolate.]

9. Khasi Mandrin

This cultivar is commercially grown in the North – East region of India. It is mainly grown as seedling trees. The cultivar produces excellent quality fruits with depressed globose bright orange, surface smooth and glossy, stalk end even or obtuse, Occasionally short necked rind thick, adherence very slight, segments 8 to 13 usually 10, pulp vesicle uniformly orange coloured coarse but melting, flavour agreeable, juiced abundant and orange colour, sweetness and acidity well blended, seeds 10 to 15, cotyledons green, polyembryonic.

10. Clementine

This cultivar is native of Algeria. The cultivar is monoembrynic, matures early and produces fruits of excellent quality.

11. Dancy

Dancy is the most important cultivar of USA. It matures mid season, it is of excellent quality, productive with a tendency to alternate bearing Dancy is closely related to ladu and keonla cultivasrs of India.

12. Beaury

It is a popular mandarin cultivar of Australia. The cultivar is similar to Dancy is closely related to Ladu and keonla cultivars of India.

13. Campeona

It is a large fruited mandarin cultivar of increasing importance in Argentina and Uruguay.

14. Ellendale

It is the principal late ripening cultivar of Australia. The fruits are large sized with attractive colour and good keeping quality.

15. Emperor

It is a leading cultivar of Australia Frit large, early mid season in maturity but quality deteriorates rapidly if stored on tree after ripening.

16. Ponkan

Ponkan is the famous and highly reputed cultivar of South China and Formosa. It is the foremost tropical mandarin cultivar, matures in mid-season, highly productive and strongly alternate in bearing.

17. Desi

It is mainly grown in Punjab and adjoining hills of Himachal Pradesh. Fruits orange colour uniform, golden yellow, rind medium thick some what thicker than coorg mandarin, segments vary between 7 and 10; pulp light reddish yellow, texture tender, sufficiently juicy, acidic but moderately flavored; seeds few, usually 3-7.

18. Darjeeling Orange

Also known as sikkim orange and is cultivated widely in and around Darjeeling hills. The trees are vigorous and prolific bearer. Fruits are relatively small in size, somewhat flat in shape colour yellowish to orange when fully ripe; rind thin, adherence little; juice plenty and sweet with good flavour ; seeds are few.

Propagation

Most of the Mandarin cultivars are propagated through seeds except kinnow and Nagpur mandarins; usual practice in coorg, Assam and North Eastern hills is to use seedlings as planting material. But with concerted efforts made to find out suitable rootstocks for different regions, orchardists hare shifted to vegetative methods, particularly T. budding because budded plants bear early, tolerant to biotic and abiotic stress. The seedling trees not only bear late but also tend to become thorny and grow tall and slender.

By Seed

For quality planting material, select uniformly matured fruits from healthy, true to type and heavy bearing plants to extract seeds. Freshly extracted seeds should be mixed with ash and dried in shade otherwise, they may loose their liability seeds are sown at a distance of 2 – 3 cm. Germination may take place with in 3 – 4 weeks. Since the seeds are polyembryonic growth are rouged out and the rest that are produced from the cells of nucleus are allowed to grow. The seedlings thus selected are more or less uniform in growth and production.

By 'T' Budding

Budding is done using the buds of bud wood taken from the disease free mother plants orn Rangpur lime, Cleopatra, Jatti khatti karna katta and Troyer citrange. Rangpur lime is a vigorous, hardy rootstock with good adaptability to a wide range of soil particularly heavy soil, tolerant to tristeza and salt; it is susceptible to footrot, exocortis and xyloporosis. Cleopatra mandarin is the most salt tolerant root stock with the ability to exclude sodium and chloride taken up by root system. It is tolerant to tristezz, exocortis and fairly tolerant to foot rot. Rough lemon, well adapted to high sandy soils. IT is susceptible to foot rot and scab and tolerant to tristeza. This is the most important rootstock for light soils Troyer citranges are used in areas where cot of hardiness and resistance to tristeza are necessary they are also resistant to foot rot but susceptible to exocortis

Karna khatta (*Citrus karma*)

It is extensively used as a root stock in North India.

Seeds of identified root stock for a particular area should be extracted from fully matured, healthy fruits. They are sown in lined (10-15cm deep) on raised seed beds inside a polyethylene house. About 1 – 2 months old seedlings are shifted to secondary beds. These are finally budded when they attain a height of 25-30cm and 1-2cm diameter. Scion should be selected from healthy, vigorous, matures, virus free and high yielding trees. They should also be free from water sprouts and chimeras.

Further use of dormant scion bud wood from past season's growth is used after it has hardened. The bud wood should be taken from round or cylindrical green twigs. T budding is done on one and a half to 2 years old seedling. In about 6-9 months, the budded plants will be ready for transplanting in the main field.

Cultivation

Planting: Generally, planting is done during monsoon in all mandarin growing areas *i.e.*, June – December. In sub mountainous tracts, where planting is generally done on slopes, proper terraces are necessary, while in plains the land should be leveled properly. Pits of 45 cubic centimeters are dug at a spacing of 6 x 6 m and filled with FYM, sand and top soil and then basins are formed. The buddlings are planted in the center of the pits and irrigated.

In N-E parts of India, Khasi mandarins are very closely spaced (4.5 x 4.5 m is ideal for kinnow budded on Jattikhatti. Kinnow can be grown successfully under high density planting by using Troyer citrange as a rootstock and by spacing the plants 1.8 x 1.8 m, accommodating 3000 lr/ha. The optimum spacing for Nagpur mandarin is 6 x 6 m when budded on Rough lemon. In Karnataka, coorg mandarin on Trifoliate orange and Rangpur lime can be planted at a distance of 5 x 5, and 6 x 6 m, accommodating 400& 275 trees/ha, respectively.

In Tamil Nadu, Mandarin are planted at a spacing of 6 x 6 m in 75 x 75 x 75cm size pits. The planting seasons are May – June and September – October.

Training and Pruning

The water shoots and rootstock sprouts should be periodically removed. Trees are trained to single stem with 4 – 6 well – spaced branches for making the basic framework. Further no branches should be allowed from the trunk up to height of 45-50 cm from the ground level. An ideal mandarin tree should be low headed with dome like crown. The bearing trees require little or no pruning. Pruning of bearing trees consists of removal of dead, diseased, criss-cross and weak branches. Removal of water shoots and suckers of rootstocks is also highly essential. Pruning of non-bearing trees can be done at any time of the year, but for bearing trees, the best time is after harvesting, during late winter or early spring when these are in somewhat dormant stage.

Root pruning is also practiced in some parts of central and southern India to regulate flowering season. However, such prunings are not beneficial in the long run.

Crop Regulation

In such and central India, mandarins bloom thrice a year. The February flowering is known as ambe bahar; June flowering as mring bahar and October flowering as hast bahar. Under such circumstances, plants give irregular and small crops at indefinite intervals. To overcome this problem and to get fruitful yield in any of the 3 flowering seasons', treating mandarin trees has been practiced which is called resting or root exposure or bahar treatment.

In this method, roots of the plant are exposed too sun by removing up to 7 -10 cm soil around 40-60 cm radius of tree trunk. The water is withheld for a month or two before flowering. As a result of water stress, leaves show wilting and fall on the ground. At this stage the roots are again covered with a mixture of soil and FYM and irrigated immediately. Subsequent irrigations are given at suitable intervals. Consequently, plants give new vegetative growth, profuse flowering and fruiting. However, in light sandy and shallow soils, exposure of roots should not be practiced and mere withholding of water for 2-3 weeks is sufficient for wilting and debilitation of trees.

It depends upon the choice of the grower as to which of the 3 bahars is to be taken to get maximum profit. As the availability of water is a problem in central India during April – May, the farmers prefer mrig bahar (June) so that the plants are forced to rest in April – May.

Resting treatment is not feasible in North India, as mandarin plants normally rest in winter and flower once a year. It is experienced that resting treatment in general is a devitalizing process and should be resorted to only under the advice and direction of a technical expert.

Manuring and Fertilization

Mandarin, like other citrus fruits also require judicious application of mineral nutrients for proper growth, development and production of quality fruits. Mandarins also require zinc, copper, manganese, iron, boron and molybdenum but not sodium and chlorine, which are rather harmful for mandarins. Improper supply of nutrients may cause serious disorders which may lead to orchard decline.

For palani Hills (Kg./tree/Yr.)

Manures & Fertilizers	*1 yr*	*II yr*	*III yr*	*IV yr*	*V yr*	*VI on wards*
FYM	10	15	25	25	25	30
N	0.1	0.2	0.3	0.4	0.5	0.6
P	0	0.08	0.12	0.16	0.16	0.2
K	0.1	0.1	0.2	0.3	0.3	0.4

For Shervaroyan hills (For trees above 6 year old)

NPK @ 700:375:600 g/tree along with VAM @ 1 kg/tree.

The fertilizers are to be applied in two splits on in May-June and another in September – October. Manures are to applied in the basin 70 cm away from the trunk at a death of 10 cm, with topsoil covered and irrigated.In hilly areas where the pH is very low, depending upon the pH, 2 – 4 kg of lime or dolomite should be applied for each tree once in 2 years, one month ahead of the application of regular fertilizers.

The spray solution containing following micronutrients can be applied once in three months at the time of new flesh production.

Zinc sulphate - 0.5%

Manganese - 0.05%

Iron - 0.25%

Magnesium - 0.5%

Boron - 0.1%

Molybdenum - 0.003%

In addition to that apply 50 g in each of $ZnSO_4$,Mn and Fe per tree per year. Application of VAM @ 20g/tree will help to accumulate Phosphorus, Zn, Cu and sulphur. The Fertigation treatment consisting of 500:240:70 NPK dose with 20% depletion of available water content found best to increase the highest conopy volume, fruit weight, TSS, Juice & yield in Nagpur mandarin.

Intercropping

eg: pea, cowpea and blackgram.

Irrigation

In south India, mandarins are grown under rainfed conditions in high rain fall areas. In winter, mandarins should be watered at 10-15 days intervals, while in summer at 5-7 days. In tarai region of UP, soil has high moisture retention capacity, thus lesser number of irrigations are required. However in Punjab, Haryana, Rajasthan and AP,more number of irrigations are required. Since root activity of mandarins is confined to a radial distance of 120 cm and to a depth of 24 cm, too much wetting should be avoided. Plants should be irrigated at 8-10 days intervals, during drought (April – June in North India and October – December in South-Central India) Mandarins are highly susceptible to water logging; therefore, stagnation of water around tree trunk should be avoided. Water should also be free of salts.

Weed Control

Weeds are a serious problem in mandarin nursery and young plantations Better way to eradicate weeds is to use weedicides. Pre-emergence application of Diuron (5kg/ha) or Terbacil (4.5kg/ha) or postemergence application of Atrazine (5-6 kg/ha) controls weeds significantly.

- ☆ **Others:** Bromocil (6 kg/ha) – Mono & dicotweeds.
- ☆ Glyphosate (5 lt/ha) – Perennial grasses
- ☆ Simazine (5 kg/ha) - Perennial grasses

Harvest and Yield

Generally, mandarins start bearing from the 4th year having 15-20 fruits/tree. However, its trees attain the level of full bearing at the age of 7-10 years. From flowering to maturity it takes 9 months.

Main harvesting periods of mandarin in different regions of India.

Region Main crop	***Off season***
North –Western plants	Dec - Feb
North –Eastern India	Nov- Feb April - May

Central India	Feb - March, October - November
South India - Coorg	December - April July - August
Nilgiris	August - October, Feb – March

In Tamil Nadu, the main season is November – December, Harvest should be done at right maturity. Therefore, fruits should be harvested when they attain full size, develop attractive colour with optimum sugar; acid blend.

Yield

About 1000 – 1500 fruits can be harvested from a tree per year and 15 –20t/ha/ year. The common practice of harvesting is to pull the fruits from the branch, which may rupture the skin near the stem-end leading to fungal infection and rotting. Therefore, fruits should neither be plucked nor torn off, but should be cut off with clippers, shears or secateurs. Although mandarins may attain optimum maturity standard but the fruits may not be attractive at the time of harvesting due to lack of good yellow colour. Accordingly, degreening of mandarins with the application of ethrel (50 ppm) one week before the harvesting develop golden yellow colour within 5 days of the treatment.

Grading and Packing

Generally mandarins and graded according to their size and appearance. Fruits are usually packed in wooden boxes for distant markets, while for local marketing; baskets of split bamboo and mulberry are used. Chopped straw and dry grasses are mostly used for padding.

The fruits should be cleaned and polished lightly with a piece of cloth, before wrapping them in tissue paper or newspaper. Use of CFB carrions in place wooden boxes is highly beneficial. Mandarins are generally transported by rail or road as ordinary cargos without refrigeration, which often leads to heavy losses due to decay and fungal infection.

Storage

Green coloured fully ripe mandarins can be stored successfully at 8 -10° C with 85 – 90% RH without impairing fruit quality. Kinnow mandarin fruit wrapped in HDPE 10G Poly bags haring 0.5% ventilation area can be safely stored up to 60 days and 80 days at ambient and cold storage respectively without much loss of quality. In Nagpur Santra, neem leaf extract @ 20% sprayed on fresh and fully matured fruits and packed in perforated polythene bags then stored in cool chamber. The results indicated that after 42 days of storage a minimum PLW (18%) & rotting (18%) Fruits (Green mature, colour break and ripe stage) dipped in 8% wax retained the freshness of fruits up to 60-3 days under ambient condition.

Acid lime: C. aurantifolia

- ☆ It is also called or sour lime. The fruit juice is rich in citric acid and ascorbic acid.

Climate and Soil Requirement

- ☆ Tropical and subtropical. Can be grown upto 1000 m above MSL. Deep well drained loamy soils are the best. They are sensitive to frost. The optimum temperature is 20 to 30°C.
- ☆ Soil pH should be 6.5 to 7.0.
- ☆ Season : December – February and June – September
- ☆ Planting: Healthy seedlings may be planted during June to December at 5 to 6 m spacing in 75 cm3 pits.
- ☆ Irrigation: Irrigated copiously after planting. After establishment, irrigation may be given at 7-10 days interval. Avoid water stagnation.

Manures and Fertilizers Per Plant

- ☆ N to be applied in two doses during March and October. FYM, P_2O_5 and K_2O are to be applied in October. Spray zinc sulphate at the rate of 0.5% (500 g/100 lit of water) thrice in a year (March, July and October) after the emergence of new flushes.

After Cultivation

- ☆ Remove branches of main stem upto 45 cm from ground level. Application of green leaves 30 kg per tree once in 3 months.

Intercropping

- ☆ Legumes and vegetable crops can be raised during prebearing age.

Growth Regulator

- ☆ To increase fruit set spraying 2,4-D-20 ppm during flowering. Fruit retention spraying- 2,4-D@ 20 ppm or NAA 30 ppm after fruit set (marble size).

CITRUS

Fruit Cracking

- ☆ It is due to sudden changes in temperature and also due to moisture stress condition.
- ☆ Cracking of fruits may be radial or transverse.
- ☆ Secondary infection is also possible due to Aspergillus, Fusarium or Alternaria Management
- ☆ Apply light irrigation at frequent intervals.
- ☆ Application of potassium during fruit development.

Granulation

- ☆ The juice vesicles become hard, enlarged and turn opaque grayish in colour.
- ☆ The density of pulp is increased, juice contains increased minerals (Calcium, sodium, potassium) and decreased carbohydrate and organic acid.

- It results in lignification of juice cells that leads to formation of sclerenchyma
- High humidity and fluctuation in temperature are the major factors.
- Young trees are more prone to granulation than older trees.
- Application of more nitrogen, excess irrigation, large size of fruits, rootstocks are also a cause.
- Mandarins on jattikhatti rootstock are more susceptible than sweet orange .

Management

- Avoid excess moisture
- Spray lime @ 20kg in 450 l of water.
- Spray zinc (0.5%) and copper (0.5%).

Sunburn or Sunscald

- The portion that is exposed to sun develops yellow patches which turn brown and become hard.
- The inner portion becomes dessicated and discoloured.
- Affected fruits are malformed and have low juice content.
- Severely affected fruits drop off and leaves turn brown.

Management

- Spraying lime solution @ 20g//l before summer.
- Regulation of irrigation to reduce the temperature.
- Mulching the tree basins.

Citrus Decline

- Also known as citrus dieback.
- Growth becomes stunted, mottling of leaves, turn yellow and are shed.
- There is excess flowering and poor fruit set.
- Affected fruits are subjected sun blotching.
- Presence of calcium carbonate or clay is harmful and leads to decline.
- Incompatibility of rootstock and scion, salinity, water logging and mismanagement of citrus orchard are causes to citrus decline.

Management

- Provide proper drainage
- Proper management of the orchard
- Use of resistant rootstocks and disease free bud wood.

ACID LIME

Plant Protection

Leaf mine

- ☆ 2 ml/l dimethoate + neem oil 3%

Leaf caterpillar

- ☆ Endosulfan – 2 ml/l when infestation is moderate to severe.

Sucking pest

- ☆ White fly : Spray quinalphos – 2 ml/lit
- ☆ Nematodes : Carbofuran – 75 g/tree
- ☆ P. fluroscens – 20 g

Diseases

- ☆ **Twig blight:** Dried twigs are pruned and sprayed with 0.3% Cu oxy chloride.
- ☆ **Scab:** Spray 1% BM
- ☆ **Tristeza virus :** Remove the infected trees and destroy. Spray monocrotophos - @ 1ml/lit to control the aphids which spread the disease. Use pre immunized acid lime seedling for planting.
- ☆ **Harvest:** Starts bearing from 3rd year after planting.
- ☆ Though harvested throughout the year, the main crop is harvested during different periods in different parts of the country. The average yield is 20-25 kg/tree/year.

Post Harvest Treatment

Treating the fruits with 4% wax emulsion followed by pre-packing in 200 guage polythene bags with 1% ventilation improves the shelf life for more than 10 days. Limes can be stored at 18°C. At HC & RI, PKM a low cost storage tank has been developed with double layer brickwork, the interspace filled with sand which is kept wet by periodical watering.

Chapter 4
Grapes

Botanical name : *Vitis vinifera* Michx.
Family : Vitaceae
Origin : Black sea to Caspian sea
Type of fruit : Berry
Edible Portion : Pericarp and Placentae
X = 19
2n = 38

SOIL, CLIMATE, VARIETIES, NUTRIENT AND WATER MANAGEMENT, INTERCULTURAL OPERATIONS

Grape is a subtropical fruit but adapted to tropical conditions. It is a vine spreading on a support, native of Armenia, a district near the caspiansea in Russia. It was introduced into India by the in values of Iran and Afghanistan. It is one of the most delicious refreshing and nourishing fruits. Fifty percent of the total production of fruits in the world is contributed by grapes. Cultivation of grapes is called Viticulture. There are about 10,000 varieties in the world.

Climate and Soil Requirement

It is a fruit of semi arid subtropical regions requiring warm dry summer and a cool rainy winter. During winter the grapes shed off their leaves and take rest. During spring they put forth new leaves and flowers. The fruits mature during summer when there is no rains. A long, warm to hot dry summer is needed for proper maturity and ripening. Grapes do not thrive in the region of humid summer as it causes fungal diseases. The distribution of rains in more important than the total amount of rainfall.

In North India heavy rains during July – September hence low production. The plant takes rest during winter since the winter is very severe and put forth new growth in summer. When the crop reach as ripening stage during june there is heavy rain resulting in poor production. In Western India grapes do not takes rest because of warm winter. Hence the grape put forth new growth twice a year. In Bombay – Deccan region and in Hyderabad the crop produced by the new lush in April maturing during rainy season (July –August) has lower market value due to less sugar content.

The climate in South India such as Bangalore in Karnataka, Dharmapuri and Madurai districts of Tamil Nadu is slightly humid and tropical. Here the maximum temperature goes upto 35°C and the minimum temperature does not fall below 12°C due to warm winter condition there is practically no rest period. Almost rainless period during November – June favours heavy as well as sweet crop. Here the vines are pruned twice. Early December pruning yields a sweet crop during April and summer pruning (May) yields a slightly sour crop during September. Theclimatic conditions are favourable is parts of Tamil Nadu so that 5 crops are taken in 2 years by staggered pruning techniques.

Soil

Well – drained rich loamy soil with pH of 6.5-7.0. Soil depth should be almost 1 m.

Propagation

Propagated by hard wood cuttings prepared from matured canes (one year old shoot) of healthy, moderately vigorous, virus tree vines. Cuttings of 25-30 cm length are prepared by making the lower cut just below a bud and upper cut slightly above the bud. Cuttings should be tied and stored in moist sand for a month for callusing. The callused cuttings start well in the nursery. While planting only one bud in left above the ground level and remaining portion buried in soil. At the end of winter the sprouted and rooted cuttings can be lifted and planted in the main yield. Grafting and budding is practiced with a particular root stock for specific requirement.

a. Phylloxera resistant root stock

   Vitis riparia, V. rupestris

b. Nematode resistant root stock

   Dogridge, Salt creek

c. Saline tolerant : Solanis, 1616

Preparation of Main Field and Planting

Trenched of 0.6 m width and 0.6 m depth are dug at a distance of 3 m apart for Muscat. Other varieties 1 m3 pits are drug. Well decomposed FYM or compost or green leaf manure has to be applied in the trenches or pit and then covered with soil. The rooted cuttings are planted during June – July.

Spacing: 3 x 2 m for Muscat

4 x 3 m for other Varieties

Irrigation

Immediately after planting 3rd day and once in a week. Irrigation with held 15 days before pruning and also 15 days before harvest.

Manuring and Fertilizer (kg per vine)

The manures should be applied twice after pruning. Apply half the dose of K immediately after pruning and the other half after 60 days of pruning. Foliar spray of 0.1% boric acid + 0.2 % $ZnSO_4$ + 1.0% urea twice before flowering and 10 days after first spray to overcome nutrient deficiency.

Special Practices

Tipping of shoots and tying of clusters in the pandal after the fruit set. Remove tendrils. Nipping the growing shoots of axillary buds and terminal buds at 12 to 15 buds. Thinning the compact bunches by removing 20% of the berries at pea stage. The clusters are dipped in a solution containing Brassinosteriod 0.5 ppm and GA_3 25 ppm at 10-12 days after fruit set to maintain vigour, yield and quality parameters.

Pests

Nematode

Carbofuran – 60 g/vine a week before pruning and irrigated profusely. The soil should not be disturbed to atleast 15 days. Application of neem cake 200 g/vine also controls nematode. We can afforded for application of *P. fluorescens*.

Flea beetles

- ✰ Phosalone – 2ml/lit after pruning and followed with 2 or 3 sprayings.
- ✰ Thrips: Dimethoate – 2 ml/lit
- ✰ Mealy bug: Monocrotophus – 2 ml/lit

Diseases

- ✰ Powdery mildew: Sulphur dusting @ 6-12 kg/ha
- ✰ Downy mildew: Spray 1% BM

Ripening

To get uniform ripening bunches are sprayed with 0.2% K chloride at 20th and 40th day after berry set and clusters of seedless varieties are diped in 25 ppm GA (25 mg/lit) at calyptras fall stage and repeated again at pepper stage to increase the size of berries.

Yield

- ✰ Seed less : 15 t/ha/yr
- ✰ Muscat : 30 t/ha/yr
- ✰ Pachadroksha: 40 t/ha/yr
- ✰ Anab-e-shahi and Arka hybrids : 20 t/ha/yr
- ✰ Grapes should be harvested only after ripening. The heat requirement of most of varieties ranges from 2900 to 3600 units. The grape berries can be kept without spoilage for 7 days at room temperature. Grapes can

economically be stored upto 40-45 days in cold storage. The optimum storage temperature recommended is -2 to -1.5°C.

☆ Raisins from grapes form an important by product industry in several grape growing countries in the world. Grapes of 17° brix and above are used for raisin making while 20-23° brix is the standard.

Chapter 5

Papaya

Botanical name : *Carica papaya* L.
Family : Caricaceae
Origin : Tropical America
Type of fruit : Berry
Edible Portion : Mesocarp
X = 9
2n = 18, 36

SOIL, CLIMATE, WATER AND NUTRIENT MANAGEMENT, PAPAIN EXTRACTION, USES, PESTS AND DISEASES MANAGEMENT

The papaya is native of Tropical America was introduced to India in the 18th century. It is now grown in almost all tropical and subtropical countries in the world.' In India, it is largely grown in Bihar, Assam, Maharashtra,. Madhya Pradesh and Andhra Pradesh with a total area of 34,000 ha, producing about 3,50,000 tonnes. Papaya is a wholesome fruit and is rich in vit-A (2000 IU/100g). Products such as jam, jelly and nectar can be prepared from the fruits. Papaya yield a valuable proteolytic enzyme, papain, which has several and varied usa in medicine and industry. Papain is used to correct certain digestive ailments, for tenderizing meat, in the manufacture of leather and in clarifying beer. The other uses of papain are in the treatment of ulcer, diphtheria, pre-shrinking of wool, manufacture of chewing gum, degumming natural silk and rayon, in cosmetics, dental paste preparation *etc*. The raw fruits are cooked as vegetables and consumed. Papaya is usually dioecious but hermophrodite type and gynodioecious types are also recognized. In dioecious type, both male and female plants are separate. The male flowers are found on long pendulous panicle. Female flowers are solitary and are much larger than male. The ovary usually large in female flowers. In the case of hermophrodite flowers, two kinds are often observed *viz*. one with long corolla type and 10 stamens and another type with a short corolla and 5 functional stamens. Fruit is a large hollow berry elongated or globular in shape. In gynodioecious type, the female and bisexual flowers are borne one on the same plant. The fruit develops from female flowers are globular in shape while fruits that develop from bisexual flowers are elongated in shape. The edible fruits are found only in *Carica papaya*. C. *candamarcensis* known as 'mountain papaya' thrives well at an elevation between 1500 to 2000m in western ghats. C. *monica* is found growing wild in Amazon basin.

Soil and Climate

It does well in varied soil types, the best performance is observed on loams of uniform texture upto 1.8 m in depth. The most important requirement is that the soil should have good drainage. Even two to three cm of water stagnation around the tree for a few hours is likely to damage them due to the collar-rot disease occurrence. Papaya performs well in tropical climates where summer temperature ranges from 35°C to 38°C. At higher elevations, the fruit quality is usually lower. It cannot tolerate very hot summer or frost, this limits cultivation in Northern India. It cannot tolerate, very hot summer or frost, a dry

Warm climate tends to increase the sweetness of the fruits. In strong wind prone areas, wind breaks have to be provided to save the trees from wind damage. Tamil Nadu is an ideal home for growing papaya because of the mild temperatures and freedom from mosaic and leaf curl virus diseases. These features help all the year round cultivation of papaya.

Seed Production

Papaya is a highly cross-pollinated crop. Seeds taken from a fruit wouldrarely breed true to type. If a variety is to be maintained pure, controlled pollination between selected female and male progenies of the same parent *i.e.*,crossing of sister and brother, called sib mating has to be done. This consists ofcollection of pollen from the male parent and applying it on the previously bagged female flower. Seeds from such sib mated fruit should be used for further multiplication. Failure to observe this precaution leads to the deterioration of the variety resulting in the progeny being a mixture of all kinds of types within a few years.

Propagation

The most common method of propagation of papaya is from seeds. Seeds are collected from well mature, ripe and large fruits borne on female plants to hermaphrodite plants as the case may be. The fruits are cut open and seeds are carefully extracted in trays. They are washed and dried in the sun or shade and are stored in bottles. Fresh seeds may be mixed with fine cold wood-ash which absorbs the slimy coating on them and helps to keep the seeds separate on drying. About 500 g seed is required for raising in one hectare. Seedlings can be raised in the raised nursery beds or in polythene bags, however the seedlings from the latter one are good. Two seeds in gynodioecious type or 5 to 6 seeds in dioecious type' should be sown per poly bag. The papaya plant can also be propagated from cuttings and grafts. Propagation from seeds is, however, preferred, because the vegetative methods of propagation are not economical.

Planting

Pits of 45cm x 45cm x 45cm size are dug at about 1.8 x1.8m apart either way. This would accommodate 3000 plants per hectare. Due to sex variations, about 40 to 60 per cent of the plant say turn to be male in the case of dioecious varieties. Therefore, in such case 2 to 3 seedlings per hole at 30 cm apart in the pit should be planted, so that when they reach the flowering phase, the unproductive male trees can be removed to keep the population ratio of one male tree for every 15 to 20 female trees. In the case of bisexual varieties, such contingency may not arise. One good seedling per pit may be planted. The best time for planting papaya is the beginning of the South- West monsoon in most parts of India. In south India, June to October and January to March are suitable for planting as the other months are either too hot or rainy.

Manures and Fertilizers

The nutrition of papaya is different from other crops because of its quick growing, continuous and heavy fruiting nature. Nutrient uptake studies conducted at TNAU showed that the uptake of N,P.K is more between flowering and harvesting stage, its peak requirement being between fruit development and harvesting. As three stages ie flowering. fruit development and harvesting concurrently occur in papaya plant, regular fertilizer application *i.e.*, 10 kg of FYM/plant as basal .besides 50g each of N, P and K per plant at bimonthly interval is recommended by TNAU. At Indian Institute of Horticulture Research. Bangalore. a dose of 250 g each of N, P_2O_5 and 500 g K_2O per plant per year in six split application recommended to get higher yield.

Irrigation

Papaya responds well to copious irrigation in well drained soils. Regular irrigation helps fruit development and induces the tree to bear larger sized fruits. Water stagnation should be avoided. In most parts of India; papaya are irrigated once in 8 or 10 days.

After Cultivation

It is not possible to identify the sex at the early stage until they put forth flowers which may take 4-5 months from planting. At this stage, male trees should be removed maintaining one male tree for every 20 female trees for proper pollination and fruit set. In each pit only vigourously growing female/hermaphrodite tree should be retained and other plants removed. During the prebearing age, short duration vegetables like cabbage, cauliflower, onion, chillies, radish, *etc.* can be grown as intercrops. Weeding should be done regularly to keep the field weed free in the young plantation as in the grown up field, the interspace remain well covered with the top growth which helps in checking weeds.

Sex Expression

Many sex forms such as dioecious, hermaphrodite, gynodioecious *etc.*,have been reported in papaya. There, are no distinct or definite methods to ascertain the sex of the plants at the early stage itself. Besides, many factors have been reported to influence the sex expression.

1) *Environment:* Low temperature tends to produce perfect flowers on the male tree and female flower production is increased in cool weather and short days. Season of planting also affects the sex expression. Planting during February shows more male plants while planting in March/April produces an equal number of staminate and pistillate plants.

2) *Growth regulators:* such as GA (50 ppm), ethrel (200ppm) SADH (250ppm) and phosphon - D (2500ppm) increase the femaleness in dioecious types.

Harvesting and Yield

The first crop of fruits becomes available in 12-14 months from the time of planting. The cropping is practically continuous during the life of the tree. In the plains of North India fruits continue to mature through the spring and summer, but in the cooler places in the hills only 3 to 4 months from February to May. Fruits should be harvested when the colour changes from green to yellowish green. It should be harvested individually with hand, taking care to avoid injuries on the fruits. The yield varies considerably and the yield per tree may vary from 50 to 100 fruits.

The yield may also vary according to the number of female and hermaphrodite trees in the orchard. TNAU bred varieties yield 100-160 t/ha. Papaya gives economic crop upto 2 years and thereafter it declines drastically. Fruits to be consumed locally should be stored in a single layer of straw until they become yellow. For distant market, it should be packed in bamboo baskets lined with straw to avoid bruising.

Extraction of Papain

The latex or'milky juice of the unripe green papaya fruit contains a largeamount of digestive enzyme called papain which is able to digest the protein in our feeds. Fully developed green large sized hard papaya fruits which are about three months old are selected for tapping. The latex is obtained by making scratches or shallow incisions on the skin of the fruit The incisions are about 0.3 cm deep. Usually not more than four incisions per fruit at equal distance are made every day. To cover the whole surface around the fruit not more than five tappings at intervals of four or five days would be necessary. Non-metallic instruments should preferably be used in tapping and collecting, as the juice acts upon metals and gets discoloured. An ivory blade or a sharp edge or piece of bamboo splinter may be used. The latex should be collected in porcelain glass or earthen containers. After about 2 to 4 hours, the latex is scraped out from the tray and dried in the sun. Tapping should be undertaken early in the morning so that drying in the sun can be done before mid-day. This makes the material sufficiently dry by the evening. When thoroughly dried, the latex becomes crisp and flaky. It may be then ground into a powder, preferably still warm. The dried papain is powdered and sieved in 10 mesh sieves. The cream coloured powder should be placed in air-tight bottles or poly bags. Papain can be also dried artificially at temperature of 50 to 55°C which will attain better colour and quality. Potassium metabisulphite (KMS) at 0.5% may be added to it for better colour and keeping quality. The papain production is influenced by certains factors such as fruit size, fruit maturity, varietal factor *etc.*

Chapter 6
Sapota

Botanical name : *Achras sapota* L.
Family : Saptoceae
Origin : Mexico (Tropical America)
Type of fruit : Berry
Edible Portion : Mesocarp
X = 13
2n = 26

SOIL, CLIMATE, NUTRIENT AND WATER MANAGEMENT, SPECIFIC PROBLEMS AND CORRECTIVE MEASURES

Sapota is a delicious fruit introduced from tropical America. It is also known as sapodilla or chiku in India. Sapota fruits are recommended to adult patients ailing from tuber culosis and children from primary complex. This offers a high economic returns even under marginal land and low input management.

Climate and Soil Requirement

Sapota being a tropical fruit crop can be grown from sea level upto 1200 M. It prefers a warm and moist weather and grows in both dry and humid areas. Coastal climate is the best suited. Areas with an annual rainfall of 1250-2500 mm are highly suitable.

The optimum temperature is between 11°C and 34°C. Sapota being a hardy tree can be grown on a wide range of soils. Soil should be well drained without any hard pan. Deep and porous soils are preferred. The most ideal soils are deep alluvium, sandy loams, red laterites and medium black soil. It can rolerate the presence of salts in the soil and irrigation water to certain extent.

Propagation

- ☆ Grafted plants on *Manilkhara hexandra* (Pala) root stock.
- ☆ Season of planting-June to December.
- ☆ **Spacing:** 8 x 8 m. High density planting 8 x 4 m.

Planting

Pits of 1m3 in drug. Filled with top soil mixed with 10 kg of FYM, 1 kg of neem cake and 100 g of lindane 1.3%.Grafts are planted in the centre of the pit with ball of earth intact. The graft joint must be alteast 15 cm above the ground level. The plants are staked to avoid bending or damage of graft joint.

Irrigation

Irrigated copiously immediately after planting and on the third day and once in 10 days after words till the graft establishes.

Intercropping: Legumes and short duration vegetable crops may be raised as intercrop during pre bearing stage.

Plant Protection

- Leaf webber: Spraying of phosalone – 2ml/lit
- Hairy caterpillars : Spraying of endosulfan – 2 ml/lit of water
- Budworm : Spray phosalone – 2ml/lit

Diseases

- Sooty mould : 1 kg maida or starch is boiled with 5 lit of water, cooled and diluted to 20 lit (5%) and sprayed.

Harvest: Mature fruits are dull brown in colour. When scratched the colour immediately below the skin will of lighter shade if matured while in the immature fruits it is green. The mature fruits are harvested by hand picking. Fruits at full maturity develop a dull orange or potato colour.

- **Season:** February – June and September – October. The fruits are ripen by keeping the fruits in a air tight chamber with. 5000 ppm Ethrel + 10 g NaOH pellets.

Yield: 20-25 t/ha/year

- Brown scaly materials disappear from the fruit surface as the fruit approaches full maturity. As the fruit matures, the milky latex content is reduced. The dried spine like stigma at the tip of the fruit falls or drops of easily when touched. Yield start from 3rd year of planting.

Chapter 7
Guava

Botanical name : *Psidium guajava* L.
Family : Myrtaceae
Origin : West Indies to Peru
Type of fruit : Berry
Edible Portion : Thalamus and pericarp
X = 11
2n = 22, 33

SOIL, CLIMATE, IRRIGATION AND NUTRIENT MANAGEMENT, NUTRIENT DEFICIENCIES, PHYSIOLOGICAL DISORDERS, PESTS AND DISEASES, MANAGEMENT PRACTICES

It is the fourth most important fruit of India in respect of area and production. It is said to have been introduced from tropical America. It is grown in many parts of the world. It is also popularly called as apple of the tropics and poor man's apple. It occupies an area of 58,000 ha in India and half of this area is confined to U.P and other important states are Bihar, Madhya Pradesh, Maharashtra, Andhra Pradesh, Kerala and Tamil Nadu. In Tamil Nadu it is largely grown in Coimbatore, Ramnad and Kanyakumari districts. It is very rich and cheap source of vitamin C (100 to 260 mg per 100 gm of the pulp) and contains a *fair* amount of calcium. It makes an excellent jelly and does not lose the vitamin C in the preserved forms. Besides *P:guajava,* the related species are:

- ✰ *P. guineense,* called Brazilian or Guinea guava which bears small fruits of poor quality .
- ✰ *P. cattleianum,* known as strawberry or cattley guava - produces very small fruits of 2.5 cm in diameter with attractive purplish red colour.
- ✰ *P. friedrichsthalianum* - (Costa Rican guava) or china guava) produces small fruits of globose in shape.

Soil and Climate

It is a hardy fruit which can be grown in poor alkaline or poorly drained soils without any manuring or irrigation. It can grow in soils with pH ranging from 4.5 to 7.5 but the best soils are deep, friable and well drained. It is a subtropical and tropical fruit which requires a distinct winter for developing good quality. It can thrive in semi arid tracts of India and enjoys cooler climate upto an elevation of 1000 m but it cannot withstand frost.

Varieties – refer Practicals

Guava varieties are generally named according to the shape, colour or smoothness of skin or from their place of origin. Varieties can be classified as seeded varieties (highly seeded to less seeded types-diploids and seedless varieties (triploids).

Propagation

In India, guava is commonly propagated from seed which germinates in about three weeks. Boiling the seeds for five minutes, soaking them in water for weeks prior to sowing or treating them in strong sulphuric acid for five minutes facilitates their germination. Propagation through this method is not desirable as the seedlings will take more time to come to bearing and seedling trees differ greatly from the mother plants. Vegetative propagation through layering is therefore recommended. Both air layering and simple layering have been found to be successful. In about 45 days, layers can be separated from the mother plants. These separate layers should be planted in full size pots and they are hardened by gradually exposing them to direct sunlight. Such hardened layers are ready for planting in about six months. Though it is hard to root semi hard wood cuttings, treating with IBA or NAA at 2000 to 5000 ppm root well under mist conditions. In some places, budding techniques using forkert, shield, patch, chip *etc* have been tried with different success.

Planting

Pits of 0:5 m x 0.5 m x 0.5 m size are dug at a spacing of 5m x 5m. The layers with the ball of earth are planted in the centre of the pit.

Manures and Fertilizers

It responds well to the application of inorganic fertilizers along with organic manures. Therefore for the bearing trees, 50 kg of FYM and one kg in each of N, P and K are applied per tree in two equal split doses, once during March and again during October. The manure and fertilizers are spread in the entire basin of the tree, 15 cm away from the trunk upto leaf drip and incorporated by shallow digging. It also responds to foliar spray of nutrients and spraying of urea 1 % + Zinc 0.5% twice a year during March and October increase the yield. Guava sometimes suffers from deficiency of micronutrients. Hence, a mixed spray containing $ZnSO_4$, $MgSO_4$. $MnSO_4$ @ 0.5% and $CUSO_4$ and $FeSO_4$ @ 0.25% plus a wetting agent @ 1 ml per 5 litre of solution at various stages *viz*. new flush, I month after first spray at flowering and at fruit set are recommended.

Irrigation

Guava though can withstand drought, it responds to irrigation at interval of 10 days.

Training and Pruning

Open centre systems or delayed open centre is generally recommended. Pruning consists of removal of suckers arising from the base of the trunk. Dried twigs and branches have to be removed and the cut ends may be applied with Bordeaux paste. The flowers are borne on the axils of current season shoots. Light annual pruning after harvesting promotes vegetative growth and flowering. In Tamil Nadu, it is recommended that the tips of 10-12 cm lengths of past seasons shoots are pruned during September and February every year to encourage more laterals. Pruned trees give large fruits and early ripening. When the trees become old, the branches are pollarded leaving 30 cm in length at their origin. The cut branches produce plenty of shoots and flowers and ultimately high yields. In the trees having upright and tall growth habits, the straight growing branches are bent and tied on the pegs driven on the ground. In the bent branches, dormant buds are activated and induced to produce flowers and fruits heavily. In certain parts of Maharashtra, root pruning is practiced to produce heavy yield. In this method roots are exposed and minute roots are cut away and irrigation is withheld so as to allow the leaves to shed. Then, the basins are covered with the manures and soil and irrigated copiously.

Cropping

The fruit buds are borne on past season growth terminally or laterally. The flowers are borne on the current season growth in the axils of leaves. The flowers are solitary or in cymes of 2 to 3. The current season growth takes one or two months to bear flowers. The floral buds require 38- 42 days for full development. Layers generally take 2-3 years for fruiting. Guava flowers twice a year, first in April-May for rainy season crop and then in August – September for wiriter season crop. In South India, there is a third crop with flowers appearing in October. As the rainy season fruits are insipid and watery and do not keep well. In certain parts of India, some practices are followed to avoid flowering and fruiting during rainy season so as to get large sized fruits of better quality during winter season. They consists of Bahar treatment - consisting of root exposure and or root pruning before the onset of monsoon.

Deblossoming of rainy season crop - spraying NAA 200-400 ppm and Withholding of water and removing the soil from around the upper roots during rainy season and covering it again with soil and manure mixture. Guava fruits should be picked immediately when it is mature and they should not be allowed to ripen in the trees lest the damage by birds and squirrels. Individual hand picking is preferable to shaking the tree. Mature or half ripe fruits are mostly prefered for consumption than ripe or over ripe fruits. Yield varies due to many factors. On an average 800 number of fruits weighing 20-25 kg may be obtained from guava. rainy season and covering it again with soil and manure mixture. Guava fruits should be picked immediately when it is mature and they should not be allowed to ripen in the trees lest the damage by birds and squirrels. Individual hand picking is preferable to shaking the tree. Mature or half ripe fruits are mostly prefered for consumption than ripe or over ripe fruits. Yield varies due to many factors. On an average 800 number of fruits weighing 20-25 kg may be obtained from guava.

Chapter 8
Pine Apple

Botanical name : *Ananas comosus* **L.**
Family : Bromeliaceae
Origin : Brazil
Type of fruit : Sorosis
Edible Portion : Bracts and perianth
X = 25
2n = 50, 75, 100

SOIL, CLIMATE, PLANTING, HIGH DENSITY PLANTING, NUTRIENT AND WATER MANAGEMENT, SPECIAL CULTURAL OPERATIONS

Varieties : Kew, Mauritius and Queen

Soil and Climate

- ☆ Mild tropical climate as found in the humid hill slopes is best suited. Can be grown in plains under shade. Elevation from 500 m to 700 m is ideal. A light well drained soil with pH 5.5 to 7.0 is preferable. Heavy soils can also be used if drainage facilities are available.

Spacing

- ☆ Plant in double rows either in beds or in trenches with the plants into the second rows set in the middle of the plants in the first row. The spacing between two trenches will be 90 cm. Row to row spacing in the same bed per trench will be 60 cm and plant spacing within the row is 30 cm.

Planting

- ☆ Use suckers and slips of 300-350 g weight for planting. Give a slanting cut to the suckers before planting and dip in Mancozeb 0.3% or Carbendazim 0.1%.
- ☆ **Season:** July – September

Manures and Fertilizers

- FYM 40-50 t/ha. N 16 g, P 4 g and K 12 g/plant in two equal splits at 6th and 12th month after planting. Apply as foliar spray 0.5%-1.0% sulphate of Zinc and Ferrous solutions at 15 days interval to overcome the deficiencies in the early crop phase.

After Cultivation

- To have uniform flowering apply the following when the crop attains 35-40 leaf stage. NAA 10 ppm + 2% urea (20 g in 1 lit of water) @ 50 ml/plant poured into crown or 2% urea + 0.04% Sodium carbonate + 20 ppm Ethephon (ethrel) @ 50 ml/plant poured into the crown. To increase the size of the fruit, 200-300 ppm NAA should be sprayed after fruit formation. To avoid calcium induced Iron chlorosis adequate shade should be given.

Plant Protection

- **Mealy bug:** Spray methyl demeton 2 ml/lit or monocrotophos 36 WSC 2 ml/lit

Crop Duration

- 18-24 months

Harvest

- Fruits can be harvested from 18 to 24 months. Slight colour change at the base of the fruit indicates maturity.

Yield

- 50 t/ha
- A plant crop and two ratoon crops are normally taken and in Mauritius variety upto five crops can be taken.

Chapter 9
Jack

Botanical name : *Artocarpus heterophyllus* L.
Family : Moraceae
Origin : India
Type of fruit : Sorosis
Edible Portion : Bracts/perianth/seeds
X = 14
2n = 56

SOIL, CLIMATE, PLANTING, HIGH DENSITY PLANTING, NUTRIENT AND WATER MANAGEMENT, SPECIAL CULTURAL OPERATIONS, PESTS AND DISEASES, MANAGEMENT PRACTICES

India is considered to be the native home of jack. In the tamil literature, jack has been given the important second position of significant 'three fruits' *viz.*, 'Mukkani'. It is mainly distributed in the tropical humid belt. In India, Assam, Bihar, Kerala and Tamil Nadu are the main jack growing states. In Tamil Nadu lower Palani hills of Dindigul Anna District and Panruti of South Arcot Vallalar District are known for the best quality jack fruits. The fleshy carpel (which is botanically the perianth) is the edible portion. Hundred gram edible portion contains 19.8g carbohydrate mainly as sugars, 1.9g protein, 0.1g fat, 1.1 g fibre, 20 mg calium, 41 mg phosphorus 0.56 mg iron, 175 mg carotene (Vit. A), 0.03 mg thiamine, 0.13 mg riboflavin, 0.4 mg niacin and 7 mg citamin C. One hundred grams of jack carpels supply 88 Kcal of energy.

Recently it has been reported that jack fruit could be very useful in the treatment of the dreaded disease of human being AIDS. An extract of jack fruit was seen to have inhibited the growth of HIV infection *in vitro*. The power of this substance called jacaline was discovered by Jean Favero, Department of Microbiology and Antibacterial limmunology, Montpellier University, France. 'Jacaline' is inactive on lymphocytes which hare already infected but has proved its might by protecting the healthy ones. After modifying this molecule to make it less toxic, scientists are planning to use in vivo (technical News from France, Centre for Documentation on Universities Science and Technology, Office of the Counsellor for Cultural,

Scientific and Technical Co- operation. Embassy of France, 2, Aurangazed Road, New Delhi, (India).

Apart from its use as a table fruit, jack is popular fruit with the housewife for making pickles, for dehydration into jack leather or thin round papad. Canned jack fruit, syrup, jam, jelly and candy have also been attempted. Preservation of ripe flackes in bottles after mixing with sugar and honey is very common in west coast. The dehydration of salted flakes for use as substitute for potato chips after frying in oil or ghee is another practice there. The outer pericarp of the fruit and sterile flowers (present in between the fleshy fertile flackes) is praised as cattle feed relished by the cattle as such or after mixing with rice gruel. From seeds, a starchy flour is made. The seeds are also relished when boiled or roasted and eaten out of hand or after soaking in syrub for some time. The seeds are also popular ingredients in many culinary preparations. The latex from bark contains a large amount of resins and is often used to plug holes in earthen containers. The timber is valuable in construction and furnishing. The leaves are sued as fodder and particularly relished by goats.

Climatic and Soil Requirements

Jack grows well and gives food yield in warm humid climate of hill slopes and hot humid climate of plains. From sea level upto an elevation of 1200 M jack can be grown successfully.

A temperature range of 22-35° C will be ideal. It can bot tolerate frost as well as drought. Under low humidity the bearing is poor and the fruits also do not develop and taste. The West coast plains with high humidity are found to be very well suited. Though it tolerates a variety of soils, a deep rich alluvial or open textured loamy soil with slightly acidic condition (pH 6.0 – 6.5) and perfect drainage is ideal.

Varieties

Cultivated jack types are classified into two groups:

(1) firm flesh

(2) soft flesh.

Singapore (or) Ceylon Jack

It was introduced in Tamil Nadu from Sri Lanka. Fruits are medium in size each weighing 7-10 kg. The carpels are crisp, sweet, yellow with strong pleasant aroma. It is a precocious bearer *viz.*, even seedling progenies will start bearing from 3 years after plantinf (normally in other types the seedlings progenies will start bearing only from 7-8 years after planting). Fruits will be available from March – June and again from September to December.

Hybrid Jack

It is a cross between Singapore jack x Veliappala developed at Fruit Research Station, Kallar. Trees are precocious in bearing; carpels are bigger in size sweeter than the parents.

Burliar – 1 : (T Nagar selection)

It was developed at Fruit Research Station, Burliar. The trees are medium in height and prolific in bearing.

PLR – 1 : (Palur-1)

It is a high yielding variety developed at Vegetable Research Station, Palur of Tamil Nadu Agricultural University. A single plant section isolated in Panikkankuppam village near Panruti of South Arcot Vallalar District of Tamil Nadu. The fully ripe fruits have flat stigmatic surface instead of a spiny surface. The special feature of this genotypes is that the trees bear fruits twice in a year *viz.*, fruits will be available in the regular jack season March to June and an off season crop during October to December is also available. Each tree bears about 60-80 fruits. The average fruit weight is 12 kg containing 115-120 flakes. The total flake weight per fruit is 2.36 kg which accounts for 19.68 % of the totalfruit weight. Flakes pale yellow in colour, crip and sweet; TSS is 190brix.

PPI – 1 : (Pechiparai – 1)

It was developed at Horticultural Research Station, Pechiparai of Tamil Nadu Agricultural University by clonal selection from Mulagummoodu local. Trees are medium tall maximum bearing in tree truck. On average each tree bears 107 fruits weighing 1818 kg per year in two seasons, *viz.*, April – June and November – December. Carpels are sweet, crisp, tasty with pleasant aroma. Suitable for commercial planting as well as for planting in home garden.

Propagation and Planting

When propagated through seeds, jack exhibits as wide variation among its progenies. Hence vegetative propagation is recommended. Commercially inarching on 10 months old jack seedlings is done to produce vegetative progenies. The grafts come to bearing within 4 ½ -5 years when compared seedlinfs which take 7-8 years normally. Soft wood grafting (cleft method) on 1 ½ months old seedlings with scion of 3-4 months old also was found to be successful. Since the viability of seeds is very low, seeds have to be sown immediately after extraction to raise rootstocks. Rudrakshi and A. hirsute are also used as rootstocks. In jack modified forkhert method of budding is also done. In Malaya air layering after etiolation is reported to be successful. At Pechiparai (tamil Nadu), patch budding on 3-5 months old rootstock seedlings done during January – February and October-November was found to be very successful. In this method the budded plants are ready for planting in 8 months. Pits of size 1 metre cube are dug at a spacing of 9-12 M pits they filled with top soil and 10 kg FYM and the grafts are planted in the centre of the pits during June – September.

Training and Manuring

Jacks are trained to single stem, early side branches should be removed then and there so that a uniform smooth trunk develops for a height of 1.5 – 2 M and then scaffold branches should be permitted to arise. This trunk is going to develop fruit buds and hence should be kept free of growths.

The fertilizers are applied during rainy season. If irrigation is available they can be split into two and applied twice in a year June – July and September – October. The manures and fertilizers can be applied in a circular trench taken 50-06 cm away from the trunk.

Irrigation

Though jack is cultivated under rainfed conditions, it is very sensitive to drought. Hence irrigation should be done depending on the type of soil, season *etc.* so that, there should not be any moisture stress especially during flowering and fruitset. Similarly too much of soil moisture will affect the quality of fruits. The flakes will develop an insipid taste when there is excess soil moisture.

Intercultivation

During the prebearing age pulses can be raised as intercrop and dried leaves can be spread below the trees to serve as a mulch for moisture conservation. As the trunk increase in size, the active buds of female inflorescence develop from the trunk as well as from main scaffold branches. This type of bearing habit is called 'cauliflorus'. These fruit buds should be protected from any possible damage by rubbing body of the stray cattle such as buffaloes through spreading dried thorny bushes. This is very essential especially when there is no proper fence in the fields.

The male inflorescence (catkins) are seen in the current season growth while female catkins are produced as cauliflorus (on trunk and main scaffold). Some times there will be more production of female flowers as well as male flowers but there will be no fruitset. This is mainly due to lack of proper pollination. The male catkins show protrusion of anthers on their surface of the spikes. The stigma becomes visible by 8 AM from 4th day after it comes from the sheath. Every day between 8.30 AM and 9.30 AM, the male catkins showing the pollen have to be collected and rubbed on the sticky surface of female of female spikes. This can be done for 10- 15 days, for each spike. Such hand pollination will help in proper fruitset and yield.

Plant Protection

Pests

Bud weevil L (*Ochyromera artocarpi*)

- ☆ It bores into the tender buds, shoots and fruits.

Management

- ☆ Destroy fallen fruits and buds, collect and kill groups, adults and then spray endosulfan (0.035%).

Diseases

Fruit rot : (*Rhizopus artocarpi*)

- ☆ It causes premature fall of young fruits due to rotting and may result in heavy loss in yield under very humid conditions.

Management

- Spraying Disthane M. 45 (0.2%0 or Bavistin (0.05%) or Fytolan (0.2%) at 15 days interval during fruit growth.

Harvest and Yield

Normally jack starts producing fruits from 7th – 8th year onwards. Grafted plants can start yield from 4th to 5th year itself. In Singapore jack even seedlings start bearing from 3rd year. Normally the fruits will be available from March to June. In higher elevation harvest extends upto September. Even in plains certain genotypes bear an off season crop during October – December. The yield ranges from 20 to 100 fruits/tree. The fruit weight varies from 10 to 30 kg.

Chapter 10
Pomegranate

Botanical name : *Punica granatum* L.
Family : Punicaceae
Origin : South east Europe to Himalayas (Iran)
Type of fruit : Balausta
Edible Portion : Juicy seed coat
X = 8,9
2n = 16, 18

SOIL, CLIMATE, PLANTING, VARIETIES, NUTRIENT AND WATER MANAGEMENT, SPECIAL CULTURAL OPERATIONS, PHYSIOLOGICAL DISORDERS, PESTS AND DISEASES, MANAGEMENT PRACTICES

Pomegranate is a favorite table fruit of tropical and subtropical countries. The aril around the seeds form the edible part which contains cool refreshing juice. Besides its use as edible fruit, pomegranate also possesses a number of medicinal properties. The juice is useful in the cure of leprosy, rind of the fruit is useful in curing dysentery and diarrhea. The colouring matter present in the fruit rind is also used in the synthesis of dyeing material for clothes. Pomegrante is native of Iran and cultivated in Spain, Morocco, Egypt, Iran, Afghabistan, Clifornia. In India though a number of states cultivate pomegranate, the main state which has the maximum area is Maharashtra. Pomegranate is a rich source of carbohydrate (14.5%), protein (1.6%), calcium (10 mg/100g), phosphorus (70mg/100g), iron (0.3 mg/100g) and vitamin C (65mg/100g)

Climatic and Soil Requirements

Despite adapted to a wide range of climatic conditions, pomegranate yields the best quality fruits in areas of cool winter and hot dry summer (which is prevalent in Baluchisthan, Afghanistan and Iran). Right from sea level upto 1850 metres, it can be successfully grown. It is fairly tolerant to low temperature, of course with differences among varieties. For proper fruit development and maturity and sweetness, a temperature of 35 – 38 0 C is needed. Under humid condition the quality gets affected. At higher elevation and areas of low temperature during winter the tree behaves as a deciduous one. It is not very specific about its soil requirement. However, in deep loamy or alluvial soil it gives very good yield. It can tolerate salinity and alkalinity in the soil to certain extent.

Cultivars and Varieties

Because of the hard seeds though the aril is pleasant, consumption of pomegranate has been a tedious and boredom process for centuries. But due to evolution of soft seeded genotypes, there is a great increase in the consumption rate of this fruit.

Hard Seeded Types

Kandhari

Fruits are large with deep red rind, aril deep blood red or deep pink with sweet, slightly acidic juice. Seeds are very hard.

Musker Red

Medium sized fruits with medium thick red rinds. Aril is fleshy with moderately sweet juice, seeds are medium hard.

Alandi or Vadki

It possesses medium sized red fruits, aril fleshy, blood red or deep pink with sweet acidic juice. Seeds are very hard.

Kabul

Large fruits, dark red with yellow parches, aril dark red fleshy seeds hard with slightly bitter juice.

Soft Seeded Types

Dholka

This cultivar has large fruits with greenish white rind, whitish to pinkish white, thick, juicy soft arils. It is the commercial variety of Gujarat.

Paper Shell

Medium sized fruits with thick rind; arils are fleshy, reddish to pink with sweet juice. Seeds are soft.

Spanish Ruby

It has medium sized fruits with thin rind , flesh rose coloured and seeds are soft.

Ganesh

Originally it was identified in Ganesh kind garden, Pune by raising OP seedlings of Alandi and designated as GBG-1. Has medium sized fruits, aril is pinkish with sweet juice. Seeds are very soft. Fruit surface smooth, yellow with red tinge, round in shape average fruit weight 325 g, TSS 16.47%, acidity, 0.42% developed at MPKVP, Rahuri, Maharshtra.

Jothi (GKVK-1)

At university of Agricultural Sciencs GKVK, Campus, Bangalore, through an evaluation of mixed OP seedlings of Bessein seedless and Dholka, a promising type GKVK-1 was selected and released. It possesses at attractive yellowish red fruit colour, medium sized fruits, red aril colour and soft seeds. Its yield potentials is 18 tonnes/ha.

Yercaud-1 (YCD-1)

At Horticultural Research Station, Yercaud, Tamil Nadu Agricultural University one superior type (ACC. No. 455) was selected with soft seeds and deep purple aril color. This was found suitable to mid elevation of Shevroys hills. The fruits are medium in size with easily peelable rind. Each tree gives 100 – 120 fruits weighing 25 kg. The average fruit weight is 350-400 g.

CO-1

It is a selection developed at Tamil Nadu Agricultural University, Coimbatore with purple aril and soft seeds.

Miridula

Developed at MPKVP, Rahuri through seedlings selection from an open pollinated F2 population raised after crosses make between Ganesh x Gul-e-Shah Red. Fruit surface smooth, dark red in colour, round in shape. Fruit weight about 250 g juice sweet, TSS 16.32%, acidity 0.47%. Seeds softer than Ganesh.

Ruby

A multiple cross hybrid developed at IIHR, Bangalore for aril colour and seed mellowness. The hybrid develops dark red arils in winter and dark pink or red aril in summer whereas in Ganesh even though the pink or dark pink aril is developed in winter, it is almost white in summer. Ruby derived certain fruit quality attributes from Ganesh, while genes for red colour of the aril was incorporated from a Russian variety 'Gulsha Rose Pink'. The fruit skin colour is reddish brown with green streaks. Rind is thin, arils are bold (37.2 g/100 arils), seed soft (2.19 kg/cm2) each fruit weighs on an average 270 g. Yield is 16 – 18 tonnes/ha.

Amlidana

It is an F1 hybrid (Ganesh x Nana) grows well under tropical climate. With quality fruit attributes Amlidana is superior to sour variety Daru whose trees come up naturally in temperate regions of North India. Its fruits provide more acidic (16.18%) 'anardana' an acidulant commercial product prepared by drying the arils of highly acidic pomegranate which is commercially marketed as condiment in North India for use in culinary preparations which serves the purpose of dried green mango (amachur) and tamarind for souring curry, chutney *etc.*

This hybrid fruits weigh 120 g each with pink bold arils. It yields 56 fruits/tree. Trees are short statured and hence suited for HDP which will give higher fruit yield/unit area.

Propagation and Planting

Pomegranate is commercially propagated by rooting of cuttings. Semi hard wood cuttings of one year old are used for rooting. The rooting is improved by application IBA 5000 ppm through quick dip method. (10 to 20 seconds). It can also be propagated by air layering or gootee, as well as ground layering. Transplanting of rooted cuttings is done during monsoon season. Pits of 60 cm x 6o cm x 60 cm size are dug at a spacing of 4 to 5 M between rows and 2 M within the rows, So that higher yield can be obtained during first 5 years. After 5 years, alternate plant within the row can be removed so as to maintain 4 x 4 M or 5x 4 M. Before planting, each pit should be filled up with 20 kg of FYM and top soil. Inoculation of 50 g of phosphobacterium+ 150 g of Vesicular arbuscular mycorrhiza in the roots at the time of planting will help better root growth and establishment.

Manuring

The following nutrient schedule is recommended for pomegranate (g/plant).

Year	N	P	K
I year	250	125	125
II Year	500	125	125
III Year	500	125	250
IV year and above	625	250	500

Besides this every year 20 kg of FYM should be applied/tree. The manures and fertilizers are mixed and applied in a round basin 1 metre away from the trunk. A week before application of fertilizer, the soil around the root zone is slightly removes and 50 g of phosphobacterium = 150 g of Vesicular Arbuscular mycorrhiza have to be applied near the feeder roots. Besides farmyard manure each tree is supplied with 10 kg of pressmud, by spreading on the top layer of soil. In sandy soils press mud serves as a mulch and prevent moisture loss during summer season. Besides this, press mud also supplies some of the nutrients.

Press mud applied plants have come to earlier bearing in sandy soil condition.

Application of 375 kg N, 1875 kg P_2O_5 and 187.5 kg K_2O through fertigation along with irrigation level at 20% wetted area is found best for pomegranate var. Mridula.

Irrigation

Drip irrigation using pitcher pot or tube will keep the soil moisture constant without much fluctuation. This helps the pomegranate to get better establishment in the early stage as well as regular bearing in the fruiting phase.

Training, Pruning and other Intercultural Operations

Pomegranate is trained as bush. Pomegranate tree has a tendency to throw out lot of suckers. If it is trained on a single stem system and if it is damaged by stem borer then the tree will be lost. Therefore, 3-4 stems are allowed per plant and they are pinched at a height of 1 metre and below the pinched tip of each stem upto 25-30 cm, 2-3 branches well distributed in all direction are encouraged. Such a training will help for good maintenance of the tree. In pomegranate the fruits are borne terminally on short spurs produced all along the slow growing mature wood. They bear fruit for 3-4 years. Every year during winter a light pruning is to be given to shorten the previous season growth so as to encourage fruiting. Besides this, dead and diseased branches, water sprouts (suckers) should be removed periodically. Water sprout from the base should be nipped at the start of their growth so a to avoid wastage of food materialin such growth which exhaust the maximum reserve food. IF such diversion of food is allowed then there will be very poor bearing in the trees. Crops regulation can be done by withdrawal of irrigation water followed by manuring and then irrigation, Water is withheld for about 2 months in advance of the normal flowering season. After 2 months, manures and fertilizers are applied and light irrigation is given. Three to four days later heavier irrigations at normal interval are followed. For this treatment the trees readily respond and produce new growth, bloom and bear a good crop.

The fruits are ready for harvest in about 5-7 months after the appearance of blossoms. Fruit cracking is a serious problem. This is mainly due to high temperature coupled with moisture stress at the time of fruit growth and maturity some times it is due to boron and potassium deficiency. The intensity increase if the matured fruits are subjected to drought or heavy rains. Cracking can be controlled by avoiding moisture stress during fruit development, application of recommended dose of 500 g of potash and bimonthly spraying of 0.25% borax = 0.1% urea during the later stages of fruit development.

Plant Protection

Pests

1. Pomegranate butterfly (or) Fruit borer. (*Deudorix isocrates*)

Infestation starts from flowering to button stage. The female lays eggs on calyx of flowers and small fruits. On hatching, caterpillars bore inside the developing fruits and feed inside. Such infested fruits may also be invaded by bacteria and fungi which cause fruit rot. Affected fruits fall down.

Management

- ☆ Spray NSKE 5% (or) neem formulations @ 2mla/1 four times at 15 days interval commencing from flowering (as oviposition deterrent)
- ☆ Release egg parasite, Trichogramma chilonis (Tricho-Cards) @ 1 lakh (16 to 20cc). Tie the tricho-card (at 4 to 5cc) per release.
- ☆ Apply endosulfan 35 EC 2 ml or dinethoate 30 EC 1.5 ml/l when the fruits are in marble size.

2. Sucking insects

Mealybug : *Ferrisia virgata*, Pseudococcus *lilacinus*. Whitefly: *Siphonimus phyllyreae*. Thrips ; *Rhipiphorathrips creutatus*, *Retithrips syriacus*. Aphids : *Aphis puniae*.

Management

- ✰ Spray 0.04% monocrophos for control of mealybugs and scale insects.
- ✰ Spray 0.03% dimethoate or phosphomidon for control of whiteflies, aphids and thrips.
- ✰ Spray kelthane (50%) 500 ml in 500 lit of water for control of red mites.
- ✰ Nicotinyl compounds *viz.*, acetamiprid 20 Sp, Imidachloprid 200 SL and thiomethoxam 70 WS can be tried. Chitin inhibitor, diaphenthiuron is also reported effective against sucking group of pests.

Diseases

1. Black spot/fruit rot (*Collectotrichum gloeosporoides*)

Symptoms

The disease starts as minute dull-violet black spots on leaves. The area surrounding the spot turns yellow, then spots enlarge and cause drying.

Fruit rot

Black pin head spot appear on the fruit at different ages. The spots will be severe on mature fruits. Black sunken spots develop and enlarge to cover larger areas of rind. The fruit rind cracks and infection spreads to interior areas and petals also. The petals become blackened and complete rotting occurs.

Management

- ✰ Spraying fruits with 0.25% mancozeb or copper oxy chloride 0.25% or carbendazim 0.1% starting from one month after flowering and repeated at monthly interval (3 sprays).

2. Bacterial leaf spot *(Xanthomonas campestris* pv. *punicae*)

Symptons

- ✰ Several minute (2-5 mm) dark coloured irregular spots surrounded by yellow tissues occur on the leaves. Later the leaves turn yellow and prematurely drop. The bacteria also attack fruits and cause dark brown irregular spots.

Management

Spraying 250 ppm streptomycin sulphate or 400 ppm streptocycline.

Harvest and Yield

The fruits are harvested when the skin turns slightly yellow and the fruit gives a metallic sound when tapped. During 4th year, the tree bears 25 – 30 fruits and a 10 year old tree gives 150– 200 fruits/year. The fruits can b stored for 15 to 20 days under ambient temperature. At 0° C and 80 % RH, they can be stored even for 2 months.

Chapter 11
Custard Apple

Botanical name : *Annona squamosa* L.
Family : Annonaceae
Origin : Tropical America
Type of fruit : Etaerio of Berries
Edible Portion : Pericarp
X = 7
2n = 14

SOIL, CLIMATE, PLANTING, VARIETIES , NUTRIENT AND WATER MANAGEMENT, SPECIAL CULTURAL OPERATIONS, PHYSIOLOGICAL DISORDERS, PESTS AND DISEASES, MANAGEMENT PRACTICES

Custard apple (*Annona squamosa*) can be called as a delicacy of dry region due to its very sweet delicate flesh. It is a deciduous or semideciduous tall woody shrub of anout 5-6 meters height having irregularly spreading branches. The fruits are rich in carbohydrate mainly in the from of sugar (23.5%), protein (1.6%), calcium (17mg/100g), phosphorus (47mg/100g) and iron (1.5mg/100g). The custard apple of India the sitaphal or sugar apple of sweet sop has many relatives.

- ☆ Bulock's heart (or) Bull's heart (or) Ramphal: (Annona reticulata) Fruits are larger in size, heart shaped, smooth and less seeded but pulp is inferior in quality.
- ☆ Sour sop: (A. muricata). The fruits have many soft spines. Fruits are highly acidic.
- ☆ Cherimoyer (or) Cherimola (or) Cherimoya (or) Lakshman phal: (A. cherimola).
- ☆ Fruits are most the delicious, slightly adicdic, sweet with buttery consistency of pulp and low seed content.
- ☆ Atemoya: A. atemoya (A. squamosa x A. cherimola). It is a F1 hybrid with a better quality of A. cherimola and adaptability to high temperature as that of A. squamosa. Besides used as a dessert fruit, custard apple can be

used in ice cream and in preparation of jam, jelly *etc*. Tropical America is considered to be the native home of all home of all Annonaceous fruits. The probable origin of A. squamosa is West Indies and South America, while A. cherimola originated in mountains of Ecuador and Peru.

Climatic and Soil Requirements

A warm climate (not very hot) with high humidity and milld winter will be the ideal climatic condition for the best growth and yield of most of the Annona sp. Except A. cherimola. Though they can tolerate extremes of temperature (beoow freezing and upto 40°C), suchextermes won't be ideal for production of flowers and fruits. They can be grown from sea level up to 1000 meteres. Where the summer temperature is very high (more than40°C) and humidity is low there willl be no fruitest though the flowering is profuse. An annual rainfall of 500-750 mm is adequate ofr growth and fruiting. Cherimoyer (A. cherimola) requires a lower temperature and a subtropical mild climate and it is difficult to be cultivated at very high temperature. Atemoyas (the hybrids between custard apple and cherimoyar) have the superior tastes of cherimoyer can tolerate fai8rly higher temperature just like A squamonsa. Ramphal (A. reticulate) does not tolerate severe summer when compared to sitaphal.

Annonas can be grown in varied soil right from heavy claly upto sandy one. They can also be grown on rocky, marginal and even waste lands. However for best yield, a well drained fertile soil with neutral pH will be ideal. The plants are shallow rooted anod hence a deep soil is not necessary. They can be grown on slightly alkaline soil and with irrigation water having slight higher pH and salinity.

Cultivars

Balanagar

It is a cultivar of A. squamosa. The fruits are greenish yellow in colour. Each fruit weight 130-140g. TSS is 20.7o brix.

Red sitaphal

It probably originated as chance seedling. Fruits of this cultivar through belong to A. squamosa are pinkish dark with erythrite red pulp. Average fruit weight is 150-160g with 22.3obix TSS.

Mammoth

It is a cultivar of *A. squamosa*. The weight of the fruit is about 125g, TSS is 200 brix.

African Pride

It is a cultivar of A atemoya (Cherimoya x sugar apple). It is a popular variety grown is subtropical region of Australia.

Pink Mannoth

It is another popular cultivar of atemoya. The pulp is similar to *cherimoyer*, being juicy with an excellent acidic flavour.

Arka Sahan

It is an *interspecific hycrid* developed at IIHR, Bangalore. Fruits are big (210g) skin is light green in colour with waxy bloom, moderately thick with large flat eyes. Fruits have improved shelf life *viz.*, take 7 days to ripe, 4 days more than 'Mammoth'. The creamy white flesh in juicy with mild pleasant aroma and tender with sparse seeds (9/100g of fruit weight). The fruit of this variety is also characterized by large segments or flakes and many of which are seedless. Flesh is very sweet (30o brix) compared to 24o brix in Mammoth. Average yield is 12 tonnes/ha.

APK (Ca)-1

It is a clonal selection from a high yielding type in State Horticultural Farm, Courtallam of Tirunelveli District of Tamilnadu developed at Regional Research Station, Aruppukkottai. It is a high yielder in rainfed vertisol (Black soil) 14.90 kg/ tree, 30.7% more than Balanagar. Each fruit weighs 207g. Average number of fruits would be 72/tree. TSS 24.5o Brix, acidity 0.2%. Adapted to semiarid plains.

Propagation and Planting

Annona is propagated commonly by seeds. Fresh seeds germinate in 20-30 days. Seed propagation results in variability in plant vigour, prolonged juvenility and inferior fruit quality. Vegetative propagation by budding or inarching on owoon seedlings and *A. reticulata* ensures genetic uniformity. Budding is usually done in early spring or in the autumn. Inarching should be carried out in early spring using one-year old mature scion and more than one year old root stock.

A spacing of 5Mx5M is recommended for annonas. At planting, 50g. phophobacterium +150g of Vesicular Arbuscular Mycorrhiza are inoculated on the roots and the plants are planted in the pits filled with 10kf of FYM and top soil. This helps in rapid growth of roots and better establishment of plants, especially in dry regions. Since Annonas are mostly cultivated on poor soils, manuring is necessary for production of good crops. Application of 10 kg Farmyard manure, 250g N, 125g P_2O_5 and 250g K_2O is recommended per tree. The fertilizers hould be applied at the commencement of rainy season. Black polythene mulch is most effective in reducing the irrigation requirement in anola cv. N.A 7 (60.86%) with an annual water requirement of 777.6 litres per tree.

Intercultivation

Intercrops like groundnut, minor millelts, crowpea and linseed can be grown in the initial years of planting. The problem of poor fruit set in custard apple can be overcome to a large extent by application of GA at a concentration of 50ppm. Dipping of freshly opened flowers in GA ensures better fruit-set, better fruit retention, increased fruit size and weight with less seed. The crop can be sprayed with NAA 20 ppm four times at weekly interval during flowering (March –April) to encourage fruitset. Irrigation and mulching during summer season also helps to prevent fruit drop.

Plant Protection

Pests

Mealy bugs (*Ferrisia virgata, Maconellicoccus hirstus)*

These cause blemishes on fruits and the pest can be controlled by spraying 0.05 per cent dichlorvos. Refer : previoue sections for newer insecticides.

Diseases

Leaf spot

Affected leaves drop down prematurely. This can be controlled by fortnightly sprays of 0.05% carabendazim commencing from the appearance of the disease symptom.

Anthracnose (*Glomerella cingulata)*

Infection begins at blossom-end of the fruit and later spreads on entire fruit surface, affected fruits shrivel and they may cling to the tree or fall down.

Management

Spray Indofil M.45 (0.02%) at 15 days interval.

Harvest and Storage

The fruts are to be harvested at correct stage of maturity. Light green fruit colour, yellowish white colour between the carpels and initiation of cracking of the skin between the carpels may be taken as maturity indices. The fruits are hand picked. The peak harvest period is October – November. A sugar apple tree usually produces 80-100 fruits per tree after 4 to 5 years.

The custard apples ripen with in a few days after harvest. The mature fruits can be stored at 15° to 20°C with RH of 85-90% and low oxygen and ethylene but with 10% CO_2. Under such storage conditions, the fruits can be kept intact for 12-18 days. Additives like Saccharified starch (1:1), high voltage treatment and packing under nitrogen gas cover, sugar (1:2) were quite effective in extending the storage of pulp at 4°C (45 days) and –18°C (90 days) temperature.

Chapter 12
Ber & Jamun

SOIL, CLIMATE, PLANTING, VARIETIES, NUTRIENT AND WATER MANAGEMENT, SPECIAL CULTURAL OPERATIONS, PHYSIOLOGICAL DISORDERS, PESTS AND DISEASES, MANAGEMENT PRACTICES

BER

Botanical name : *Ziziphus mauritiana* Lam.
Family : Rhamnaceae
Origin : India
Type of fruit : Drupe
X = 12
2n = 24, 48, 96

The ber is a vigorous growing, small spreading tree with almost vine like drooping Branches. The round to oval reddish brown are having 5.4-8.0% sugar and 85-95 mg of Ascorbic acid (Vit. C) per 100g. Central Asia is supposed to be the centre of origin for ber. The tree is a host plant for rearing lac insect (Tachardia laccad). Lac insect rearing helps in the production of lac. The powder of ber roots has very many mewdicinal properties such as cure for ulcer, fever and wounds. The stem bark powder is a remedy for diarrhea.

Climatic and Soil Requirements

India ber Zizyphus mauritiana is to grow in tropical and subtropical regions while, zizyphus jujube, the Chinese ber is a deciduous tree found in temperate region. Ber is an ideal fruit tree for arid and semiarid regions where copious irrigation is impossible. It can tolerate a high temperature of even 40°C. But below freezing temperature is injurious. Since it possesses a deep tap root system, it can be grown in an extreme moisture stress and in a wide range of soil upto an infertile one which is unsuitable for major fruits and other crops. In alkaline soils with high ph (even upto 9.5) and sodic soil, five kg of gypsum has to be applied pr pit, mixed with top soil and flooding the pits one weak before planting has to be done. By such a practice, the ber plant can be established. Once it gets established. It tolerates salinity even to an extent of 21m. mhos per cm. Z. jujuba can be used even for biological reclamation of saline soils. To certain extent ber tolerates water stagnation too.

Varieties

Kaithili

It is a variety with straight thorns but not so pronounced. Leaves are ovate with minutely serrated margin. Fruits ovate-oblong with broadly mummillate apex, 3.37 cm long, 1.9 cm thick weighing 6.22 g. stone elliptic oblong with pointed tip and furrowed surface.

Umran

In this variety, the trees are medium sized with bushy decumbent branches almost touching the ground. The thorn is curved. Ovate oblong leaves with prominent serrations. Fruit elliptic, 4.2 cm long and 3.2 cm thick.

Gola

It has got spreading tree. Fruits are almost round with flat stylar end. Skin is bright yellow, smooth and glossy, fruits come to ripening during January. Each fruit weighs 14-25g. each tree yields about 100-125kg.

Seo (Sanaur No.2)

leaves ovate to ovate oblong with obtuse base and acute apex. Fruits round resembling crab apple, stylar end round with mild depression in the centre, stem end broad, deeply grooved. Fruits light pinkish yellow with occasional specks at maturity. Selected from a place called Sanuar near Patiala in Punjab.

Seb

It is an early variety. Fruits are golden yellow in colour and slightly oblong *i.e.*, 3.0cmx 2.5cm. it yields 90-1000 kg per tree. It acts as a good pollinizer for a number of varieties.

Banarsi

It is a mid-season variety. Trees are 8-12 M tall. fruits globose oblong to long in shape with tapering stylar end. Unripe fruits are green in colour. After ripening they turn to golden yellow.

Chhuhara

It is another mid-season variety with semi-tall tree having spreading branches. Fruits ovate-oblong, size 2.9 cm x2.1cm; weight 16.8g. fully matured fruits which start ripening will be greenish yellow in colour. After full ripening, the colour changes to chocolate brown and the skin becomes very thin and soft. The flesh will be very sweet. Fruits are suitable to be heated and made into dry fruits like dates.

Sandhura Narnaul (Sanaur No.1)

It has erect trees. Fruits are oval-oblong to longish, stylar end slightly pointed. Fruits are greenish yellow to golden yellow. Size 4.45 x 2.18 cm. it has thin skin. The average yield is 80 kg/tree/year.

Elaichi

Trees spreading with fruits having the characteristic shape of cardamom hence called 'elaichi'. Fruits are small each weighing 6g with thesize of 2.05 cm x 1.88cm. the average yield is 115kg/tree/year.

Propagation, Preparatory Cultivation and Planting

Ber is propagated by 'T' budding or inverted 'T' budding on seedling rootstock of *Z.jujuba, Z. xylocarpa* and *Z. rotundifolia*. Fruits of wild species are taken; seeds are extracted and soaked in 17% salt solution to remove the ill filled ones which float. The seeds which sink are taken and soaked in con. H_2SO_4 for 5 minutes washed and soaked for 48 hours in cold water. Then the seeds can be sown in the poly bags (25 x 15 cm) of 300 gauge thickness. The seeds take 10-15 days for germination. Since the development tap root system is very fast in ber as well as in wild species, when the seedlings are with two leaves they are transplanted in the main field in pits of size 1x1x1 m filled with 20 kg of FYM + top soil and irrigated. The treated seeds can also be sown directly in to the pits @ 2-3 seeds per pit at a depth of 3 cm. normally the required varieties are budded *in situ* on this rootstock seedlings after 90 days.

If we want to bud the seedlings raised in polybags, large sized polybags have to be used, since the tap root grows very fast. June to august is the best period for budding for getting maximum bud-take. The scion sticks with 0.9 cm dia, about 1 year old maturity with plumpy buds should be selected to take buds. The buds will take about 7-10 days for sprouting.

Training

For young plant, a support should be provided by bamboo stakes. In the first year, all the branches arising upto 75 cm-1M should be removed so that a straight trunk can be developed. Above this 3-5 primary branches should be developed so as to have a balanced frame work in all directions. In the second year, on each primary branch, 3-4 well distributed secondary branches ca be allowed, and during third year final frame work should be decided.

Manures and Manuring

The plant should be fertilized in the following ratio at every year.

Age	FYM (kg/tree)	N	P	K
1 Year after planting	10	125	40	75
2 year after planting	15	250	80	150
3 year after planting	20	250	120	225
4 year after planting and then onwards	25	500	160	300

The manorial dose can be split into two equal halves and applied once during June and another after pruning. The manure fertilizer mixer should be placed in trenches of 30 cm width formed at 1.0-1.2 M away from trunk.

Irrigation

For young trees, irrigation should be done once in a work. As the trees grow older, it can restricted once in 15 to 20 days. At the time of flowering and fruiting, there should not be any moisture stress. In rainfed condition, rain water harvest can be done by forming a saucer basin giving a 0.1% slope towards the trunk of the tree from a distance of 3 M.

Interculture and Pruning

The interpace can be utilized to raise a crop like pulses such as greengram, blackgram, cowpea *etc.*, for about 3 years. If sufficient irrigation facilities are available papaya or phalsa can be grown as inter crop for first three years. In ber, the fruit buds are borne on current season growth in the leaf axils. Therefore the plants should be pruned every year to induce new growth to provide maximum fruit bearing area. After the harvest of fruits, the plants have to be pruned by heading back 25% of one year old shoot. If severe pruning is attempted, it will adversely affect the growth leading to poor yield. Dry, dead, diseased wood and criss-cross branches should be removed.

Plant Protection

Pests

1. Fruit fly (*Carpomyia vesuviana*)

The flies puncture the young developing fruits by inserting their ovipositor and lays eggs singly. Hatching of eggs completes in two to three days. The larvae (maggots) feed inside the fruit pulp and make small holes in the rind and come out of the fruit when fully grown. The affected fruits become misshapen and their growth is retarded.

Management

- ✰ Collect infested fruits and destroy them
- ✰ Spray monocrotophos(0.04%) or Rogar 30EC(0.06%).
- ✰ Spray with a mixture of 100 ml Malthion 50EC and one kg jaggery or sugar in 100 lit of water twice starting from September to October at 7-10 days interval (3 sprays).
- ✰ Cultivate ber orchard soil during April-May and apply 10% BHC to destroy pupae.
- ✰ Grow fruitfly tolerant varieties like Umran, Sanaur

2. Bark eating caterpillar (*Inderbela quadrinotata*)

The caterpillar maks holes in the trunk while feeding. Affected trees become stunted and yield potential is reduced.

Management

- ✫ Remove frassy galleries and paint the bark with 0.05% monocrotophos 40EC.

3. Hairy caterpillar (*Euproctis freterna*)

The caterpillar feed on leaves and cause damage.

Management:

- ✫ Dust 10% BHC.
- ✫ Spray carbaryl at 0.15% a.i.

Ber beetle or leaf chafer (*Adorentus pallers*)

Beetles feed on leaves mainly during night. The leaves become just like a sieve.

Management

- ✫ Spray with one kg carbaryl in 300 litres of water.
- ✫ Use light traps.

Diseases

Powdery mildew (*Oidium emysiphoides*)

Affected fruits show white powdery spots which later cover whole area of fruits. The white powdery mass also spread on flowers and leaves. Later white spots turn brown and fruits drop.

Management

- ✫ Spray dinocep 0.1% at 15 days interval after fruit set preferably at pea stage.

Harvest and Yield

In ber, fruits harvested at correct stage of maturity alone will ripe properly. Aftr attainment of full size of a particular cultivar and turning of colour to yellow or golden yellow, the normal harvesting season is October-November, while in North India it varies from place to place *viz.*, December to April. The average yild from a 10-20 year old tre would be 100-200 kg/year. If the fruits have to be stored, they can be stored at 3°C and 85-90% for 30 to 40 days.

JAMUN

Botanical name : *Syzygium cumini Skeel*
Family : Myrtaceae
Origin : East India-Malaya
Type of fruit : Drupe
Edible Portion : Epicarp and mesocarp
X = 20
2n = 40

It is a tall handsome evergreen tree of tropical and subtropical regions and has its origin probably India or East Indies. It is one of the most hardy fruits and best suited for wastelands. It is drought tolerant, at the same time can tolerate water stagnation and marshlands, where other fruit crops can not be grown successfully. The wood is used as a timber in building and railway sleeper. One hundred gram of fruits contain 19.7 g carbohydrate, 0.7 g protein, 1.0 g iron, 0.02 f calcium, 0.01 g phorus 0.1 g fat and 0.9 g fibre. Besides taken as a dessert fruit, it is also used to make beverages, squash, jam, jelly and wine. Fruit syrup is used in curing diarrhea.

Climatic and Soil Requirements

It is adapted to tropical and subtropical conditions requiring a dry climate during flowering and fruiting. Early rains resulting in better soil moisture will help in growth anddevelopment and ripening of fruits. Well drained deep loam would be ideal though it can grow on a wide range of soil conditions. In the initial stages of establishment there should not be any drought. When the tree has grown sufficiently it can tolerate drought as well as flooding.

Cultivars

In North India a cultivar known as 'Ra Jamun' with big sized fruits is being cultivated. One seedless type (with under developed ovule) was isolated at Horticultural Research Station, Periyakulam is being grown in Agricultural Research Station, Paramakudi. But the fruits are very small.

Propagation and Planting

Though it is propagated by seeds, for true to type progenies vegetative propagation is recommended. Air layering and application of 1000 ppm NAA or IBA as lanolin paste can give 80 – 90% rooting. Inarching, veneer grafting and forkert method of bedding are also used. The plants are planted at a spacing of 10-12 m in pits of 1 cubic metre size.

Pruning, Manuring and Irrigation

The plants should be trained initially in such a way that the branches develop from a point 100 cm above the ground level. Only during the initial 2 – 3 years the plants require regular irrigation. In later years the irrigation is required mainly during fruit growth and ripening. Only in poor soil the plants require manuring, 20 kg of FYM at prebearing and 80 kg at bearing age per tree is recommended. When the soil is fertile the manuring and irrigation are withheld to encourage fruit bud production.

Flowering and Fruiting

The flowering starts in first week of March and continues upto April end. The pollination is done by honeybee, houseflies and wind. By natural fruit drop only 12 – 15% of flowers reach maturity. Excessive fruit drop can be reduced by two sprays of 60 ppm GA, one at full bloom and second 15 days after fruit set.

Common Insect Pests

Fruit fly *Bactrocera (Dacus) correctus*

Leaf caterpillar *Carea angulata*

Purple winged moth *Bombytelia delatrix*

Psyllid Trioza jambolanae

Thrips *Leeuswenia ramakrishnae*

White fly *Dialeurodes eugeniae*

Plant Protection

Leaf eating caterpillars can be controlled by spraying dimethoate/malathion. White fly damages all parts and even fruits get wormy. Affected, dropped fruits should be collected and burnt. At the time of flowering if spraying has to be taken up only a safe insecticide to honeybees like endosulfan has to be sprayed. Leaf spot and fruit spot caused by *Glomerella* can be controller by Indofil Z. 78 (2 g/lit.)

Harvest and Yield

The seedling trees start bearing after 10 years, while the vegetatively propagated progenies come to bearing in 5- 6 years. The fully ripe fruits should be picked by hand by climbing on the trees and collecting in a bag. Since the jamun fruits are highly perishable and hardly stand only 5 days, they should be immediately sent to market. However if necessary they can be stored for 3 weeks at 9° C and 85 – 90% RH. They yield ranges from 70 – 100 kg/tree/year.

Chapter 13
Aonla

Botanical name : *Emblica-officinalis* (Syn. *Phyllanthus emblica*) Geartn.
Family : Euphorbiaceae
Origin : Tropical Asia (Indo-China)
Type of fruit : Berry
Edible Portion : Mesocarp and Endocarp
X =7
2n =28

SOIL, CLIMATE, PLANTING, VARIETIES, NUTRIENT AND WATER MANAGEMENT, SPECIAL CULTURAL OPERATIONS, PHYSIOLOGICAL DISORDERS, PESTS AND DISEASES, MANAGEMENT PRACTICES

Being a member of Euphorbiaceae to which most of the xerophytes, the cacti and succulents belong, aonla is a hardy drought resistant fruit tree. A rare combination of character is its ability to withstand water stagnation too. It is also known as Aonla, amali and nelli *etc*. The Aonla fruits are a rich source of Vitamin C. The special attribute is its capacity to retain Vitamin C even in a dried state which is not possible in other fruits. The vitamin C supplied by its fruits and dried powder is even superior to synthetic Vitamin C. One part or other is used in the cure of cough, bronchitis, jaundice, diabetes, dyspepsia, diarrhea and fever. Hundred grams of fruit pulp contains 14 g of carbohydrate, 0.5 g protein, 1.2 g iron, 0.3 mg vitamin B and 600 mg of Vitamin C. Because of the high content of iron (1.2mg/100g) and B vitamin C. Probably this fruit extract is used in a number of ayurvedic and homeopathic preparations which are said to prevent greying of hairs and falling of hairs. The probable centres of origin are the South and Central India, Sri Lanka, Malaysia and South China.

Climatic and Soil Requirements

Though Aonla is a subtropical fruit, it thrives very well and comes to yield in tropical humid conditions also. If young plants are protected from both extremes of temperature, the mature trees can tolerate right from freezing (0°C) to as high as 460C. Similarly if the young plants are protected from severe drought during summer and water stagnation during heavy rains, then the adult trees tolerate these two adverse conditions.

Though in a well-drained loamy soil the Aonla trees grow faster, they can grow in a heavy clay also provided slight drainage is arranged during heavy rains in their early crop growth phase of first 2 – 3 years. They also fairly tolerate alkalinity in soils (pH 8.5) and irrigation water.

Cultivars and Varieties

Chakaiya

It is a prolific bearer. The fruits are flat, with smooth skin which is greenish in colour. Fruits are small to medium in size weighing 26 g having a TSS of 10.70 brix. The trees are spreading in nature.

Banarasi

The fruits are large in size flattened oblong with smooth skin, yellowish with characteristic three raised segments. On an average each fruit weighs 38g. It is a shy bearer. Trees are having upright growth habit.

Krishna (NA – 4)

A chance seedling of Banarasi developed at Narendra Deva Agrl. University, Faizabad with medium to large size (40 g) conical, angular, smooth yellowish fruits with red blush on the exposed surface. It has fibreless flesh which is semitransparent and hard. It is a moderate bearer.

Kanchan (NA-5)

It is supposed to be a chance seedling of Chakaiya. A profuse bearer with small to medium-sized fruits (32g) flattened oblong; skin is smooth, yellowish in colour. It was also developed at Narendra Deva Agrl. University, Faizabad.

Francis

Fruits are large (41.5g) with a TSS of 12.00 brix, flattened oval, greenish yellow in colour, soft and almost fibreless. The branches have dropping habit.

BSR-1

It was developed at Agricultural Research Station, Bhavanisagar of Tamil Nadu

Agricultural University, Tamil Nadu. The trees are medium in height and spreading and hence suited for accommodating more number of trees per unit area (can be planted at 6M x 6M spacing). Fruits are medium in size (27g) with flat stylar end and round calyx end. Each fruit has 6 segments, the juice has a TSS of 18.10 brix.

Propagation

Seed propagation which has been in practice has given lot of variation in the progenies. Hence vegetative propagation was resorted to. Building using 1 year old rootstocks through "T" method (shield) or patch method is successful. In-*situ* budding will be better than budding in nursery and transplanting the budded plants. To raise rootstock seedlings, from the fully ripe mechanically or by drying in sun and the seeds are extracted. The seeds are hard and take long time to germinate. Hence the seeds should be treated with con. H_2SO_4 for 3 minutes and then washed with water and soaked in 500 ppm of Gibberellic Acid for 24 hours. Such treated seeds can be sown in nursery bed or polybag filled with pot mixture.

Field Preparation and Planting

After deep and thorough ploughing, pits of 1 cubic metre size can be dug at a spacing of 9 Mx9M during May – June and filled up with 10-15 kg of well decomposed FYM. The budded plant or seedlings for in-situ budding (especially in dry areas) can be planted at the onset monsoon rains.

Training, Pruning, Intercultivation and Irrigation

In training the trees, initially all the rootstock sprouts should be nipped off then and there. The growth of scion should be trained to develop a low headed one. At about 75 cm to 1 M height the first two main laterals are allowed to grow out. Five to six such laterals with wide angled crotches should be encouraged at fairly well spaced on the truck so as to have a modified leaser fashion.

When the crops started bearing early year after the harvest of fruits, dead, diseased, weak and criss-cross branches should be pruned. The water sprouts and rootstock growth should be watched and periodically removed. During early stages of establishment, the plants should be watered periodically especially during summer. Fully mature trees are seldom watered. However, irrigation during April-June one in 15 days will help to encourage fruitset and prevent fruit drop. During early stages of crop growth an inter crops like cowpea, greengram can be grown upto 4 – 5 years. Black polythene mulch is most effective in reducing the irrigation requirement in aonla cv. N.A7 (60.86%) with an annual water requirement of 777.6 litres per tree.

Manures and Manuring

At the end of first year 20 kg FYM and 30 g N should be applied. An additional quantity of 30 g of N should be added eary year as the tree grows. For a bearing tree, 800 g N, 640 gP_2O_5, 750 g K_2O should be applied along with 30 – 40 kg of FYM every year during rainy season. If irrigation facilities are available, this dose can be split into two equal halves, one half applied during April – May (at fruit set) and irrigated. The other half can be given during September – October.

Common Insect Pests

Aphid *Setaphis bougainisillia*

White fly *Trialeurodes rara*

Bug *Scutellera nobilis*

Leaf roller *Caloptila acidula*

Plant Protection

Bark borer makes tunnels along the bark. This can be controlled by injecting kerosene oil in the holes and plugging with cotton and clay during September – October and February – March and spraying with phosphomidon 0.03% Shoot gall maker can be controlled by pruning the affected shoots and spraying 0.1% monocrotophos. Aonla rust can be controlled by spraying dithane Z.78 or M.45 at 0.2%. blue mould develops water soaked lesions on fruits which is ultimately covered by bluish green pustules. This can be checked by a weak solution of borax or sodium chloride.

Harvest and Yield

The trees will start yielding from 8 – 10 years after planting. Grafted or budded plants will yield from 5th year or 6th year. The average yield is about 150 kg/tree/year.

Chapter 14
Wood Apple & Bael

SOIL, CLIMATE, PLANTING, VARIETIES, NUTRIENT AND WATER MANAGEMENT, SPECIAL CULTURAL OPERATIONS, PHYSIOLOGICAL DISORDERS, PESTS AND DISEASES, MANAGEMENT PRACTICES

WOOD APPLE

Botanical name : *Feronia limonia* L.
Family : Rutaceae
Origin : India
Type of fruit : Amphisarca (Berry)
Edible Portion : Succulent plancentae
X = 9
2n = 18

Wood apple, a native of India and Sri Lanka is one of the hardy trees of arid and semi arid regions. The fruit is a hard- shelled many seeded berry with its pinkish brown aromatic soursweet pulp being the edible portion the seeds embedded in it. The pulp contains 18.1% carbohydrate, 7.1% protein, 3.7%, fat, 5.0%, fibre and 1.9% mineral matter. The pulp is a rich source of calcium (130 mg/100g), phosphorus (110mg/100g) and iron (0.48 mg/100g). the vitamins supplied by one hundred gram of pulp are carotene 61 µg, riboflavin 0.17 mg, niacin 0.8 mg, thiamine 0.04 mg and vitamin C 3 mg. the ripe fruit pulp makes excellent chutney and it is also consumed afresh along with sugar. It is used as an adjunct in jelly preparation along with the pulp of guava.

Climatic and Soil Requirements

Wood apple can be grown in dry tracts of tropical and sub-tropical regions right from sea level, upto 1400 m above MSL. It is adapted to a wide range of soil conditions including degraded soil. It can also tolerate salinity to certain extent. It is an ideal tree to be exploited for growing in wasteland.

Cultivars and Propagation

No named cultivars are available. However sour and sweet types, high yielders with big fruit size exist in the variable seedling progenies in nature. High yielders with big sized and sweet tasting fruits should be selected for propagation through vegetative means. At present though seed propagation is done mainly, budding has been reported to be successful if done during late summer and early monsoon. Budded plants are dwarf and precocious in bearing. Indry regions where irrigation potential is limited. The seedlings can be planted in the field and *in situ* budding has to be done on established seedling.

Field Preparation and Planting

Normally wood apple is not planted in fertile or rich soils. In wasteland, if mass planting is to be done, then pit lines are drawn across the slope and pits can be dug at a spacing of 8Mx8M each pit with a size of 1 Mx1Mx1M. Planting should be done at the onset of monsoon after filling the pit with 20 kg FYM, sand and top soil. The basins should be formed immediately after planting in such a way that water harvesting is facilitated.

Interculture

Training is done by Central leader method allowing well spaced branches in all directions. Intercrops can be taken during rainy seasons for the first 5 years. In the post monsoon season, the basins can be mulched with dry leaves. Every year 25 kg of FYM is to be applied for each tree at the beginning of the monsoon rains. This will help in increasing fruit- size and quality. During early stages of crop growth, if pot watering is done during summer it will be beneficial. Being a member of citrus family it is attacked by the leaf-eating caterpillar of citrus which completely defoliate the plant. Spraying of any contact insecticide should be done after hand picking and destruction of larvae.

Harvest and Yield

Budded plants come to bearing 3-4 years after planting. But to reach optimum productivity it will take about 10 years. The crop flowers in February to May depending on the climatic conditions of a locality and fruits will be available from July to December depending on the flowering month. A well grown tree will give 200-250 fruits/year.

Bael

Botanical name : *Aegle marmelos*
Family : Rutaceae
Origin : India
Type of fruit : Amphisarica (Berry)
Edible Portion : Succulent
X = 9
2n = 18

Beal, one of the oldest fruits cultivated in India has a mythological significance *viz.*, a sacred tree whose leaves are used for worship of Lord Shiva. The fruit pulp which is carbohydrate. One hundred gram of pulp contains 55 µg of carotene, 0.13 mg thiamine, 1.19 mg riboflavin, 1.1 mg niacin and 8 mg vitamin C. from its pulp sherbet and syrup can be made. The marmalade prepared from its fruits is used in curing diarrhea and dysentery. From the stem, gum is obtained. The wood is used for making agricultural implements. The leaves are used as fodder. All parts of the plant are medicinally important due to a substance called 'marmelosin'.

Climatic and Soil Requirements

A subtropical condition with hot dry summer and mild winter would be ideal for the cultivation of bael. It can be grown even upto an altitude of 1200 M MSL and it is not damaged by temperature even as low as -7° C. Since it is a hardy tree, it thrives well in a wide range of soil right from pH 5 upto pH 10 where many other fruit trees fail. it can tolerate even every alkaline soil as well as stony soils. however, well drained sandy loam is the best.

Cultivars and Propagation

There are lot of variation seen among the progenies raised from seeds for size and shape of fruits, bearing habit, pulp quality, colour, texture, sugar percentage *etc.*, Cultivars like 'Mirzapuri', 'Kaghli', 'Gonda' and a few selections from Faizabad like KB 11, KB 1, Dhar Road and Ayodhya are found to be better. Root stocks are raised from seeds. On 6 months old seedlings, patch budding is done during June-July.

Planting and Interculture

Pits of size 50 cm x 50 cm x 50 m are dug at spacing of 10 M x M. The top soil should be mixed with 10 15 kg of FYM and pits filled up. Planting can be done during June-July or at beginning of monsoon. Every year regular application of 20-30 kg FYM should be done at the beginning of monsoon. Young plants are irrigated whenever there is monsoon failure. Any legume or forage crop can be taken as inter crop during early years. Young plants are trained with the help of stake. No annual pruning. However, criss-cross, weak and broken branches have to be removed periodically.

Harvest and Yield

Seedlings take 7-8 years for bearing, while budded plants start bearing at the age of 4-5 years. Flowering is seen in May-June and fruits become ready in 8-10 months *viz.*, April May matured fruits to be harvested individually along with fruit stalk and they should not be allowed to fall on the ground. A well grown tree of about 12-15 years age gives 300-500 fruits/year.

Chapter 15
Date Palm

Botanical name : *Phoenix dactylifera* L.
Family : Arecaeae
Origin : West Asia and Arabia
Type of fruit : Drupe
Edible Portion : Pericarp
X = 18
2n = 36

SOIL, CLIMATE, PLANTING, VARIETIES, NUTRIENT AND WATER MANAGEMENT, SPECIAL CULTURAL OPERATIONS, PHYSIOLOGICAL DISORDERS, PESTS AND DISEASES, MANAGEMENT PRACTICES

Date palm is a nutritive fruit rich in sugar and iron and predominantly seen in desert oasis. It is believed to have originated in countries around persian gulf such as Iraq (Mesapatomia) and Egypt. The flesh contains 60 – 65 percent sugar, 2 percent protein, 2.5 per cent fibre, 0.4 per cent fat and 2 per cent mineral matter. It is a good source of easily assimilable iron (7.3 mg/100g) and hence prescribed to anemic patients along with honey.

Climatic and Soil Requirements

Dates can be successfully cultivated in areas having long hot summer and mild temperature during flowering (February to April) and fruit ripening (May to August) should be 25 to 29°C. As in the case of grapes, this crop also requires specific heat units (above a base of 10°C nearly 3000 units for successful fruit maturation. Three should not be any rains or high humidity during fruit maturity and ripening. Otherwise the fruits will be spoiled. The crop is susceptible to frost.

Deep sandy loam are the best suited though it is very hardy and can be grown in a wide range of soil conditions. Better water holding capacity with good drainage is desirable. It can grow in alkaline and saline soil, however, the growth and productivity are affected.

Cultivars

Depending on the season of ripening, the cultivars are classified as early, mid and late cultivars. Though there are nearly 40 cultivars imported from Middle East and North Africa, few only were found to be promising, under Indian conditions.

Halawy

An early variety with small fruits. At full maturity before ripening (doka stage) the fruits are yellow in colour and free from astringency.

Khadrawy

Small to medium sized fruits; light yellow at doka stage.

Barhee

Mid –season to slightly late cultivar with small to medium sized fruits, yellow colour and low astringency at doka stage.

Zahid

Mid – season cultivar with small to medium sized fruits ; it is tolerant to rain and humidity.

Medjool

A late cultivar; fruits are large, broad, oblong-ovate, orange yellow with reddish brown stippling at doka stage.

Propagation and Planting

Though it is propagated by seeds, the seedlings exhibit a very long juvenile phase nearly taking 7 – 8 years for flowering. Besides this, the population segregates for male and female plants and the resulting all female plants also may not be uniform in bearing. Hence vegetative propagation through off-shoots produced by a female mother plant of particular cultivar can be used for planting . After about 4- 6 years, 2-3 off shoots can be obtained from a mother plam for a period of 8 – 10 years. To encourage rooting, the base of off-shoots arising from the mother plant are applied with moist soil by putting a soil boxes at least for a period of one year before separation. Each matured off-shoot may weigh 25-35 kg. Planting of off-shoots can be done during late spring or early summer by cutting with a help of a specially designed chisel. Most of the basal leaves are removed and only the terminal bud with 10-12 leaves around it are retained. Depending on the soil fertility, the spacing varies from 4M – 9M. Commercially as spacing of 6 to 8 M is adopted. In India a spacing of 3 M – 4M is adopted. For effecting pollination under commercial cultivation 2-3 male plants are planted for every 100 female plants.

Irrigation, Manuring and Interculture

Light and frequent irrigations have to be given immediately after planting and the basins can be mulched. Though date palm is drought tolerant, 2-3 M around root zone is to be kept moist for maximum growth. The irrigation frequency varies with season. Date palm is highly tolerant to saline water (even upto 2500 ppm). But to overcome drainage problem arising out of saline water usage, periodical leaching with food water should be done. Each palm should be applied with 50 kg of FYM, 600 g N, 100 g p and 700 g K. In date palm older leaves which are in surplus than required are normally pruned. To get optimum yield with food palm are retained. The leaf pruning is done during June. By this pruning, the bunches will be better ventilated during July and early August, which will be otherwise spoiled. The spines from the leaves around the bunches are also cut during spring to facilitate pollination.

Hand pollination is done by inserting 2-3 strands of male flowers between strands of female flowers when the female spathes have crack open. Though the dried pollen can be stored at 4-5°C until next season, the fresh pollen produces the best fruit. The spathes emerge during February – March and flower opening starts during March – April. Immediately, the flowers should be pollinated (2-3 days after spathes open). Pollen grains of certain varieties can advance ripening of certain varieties. So specific polliniser varieties should be identified for specific varieties of female plantation.

Fruitset

Fruit thinning has to be resorted to so as to retain 1300 – 1600 fruits in 8-10 bunches per palm will be optimum. Bunch thinning can be done either by removal of entire strand or shortening of strands. Spraying ethephon @ 200 ppm 10 – 30 days after fruitset will help to thin fruits effectively. It also helps to overcome biennial bearing and encourage earlier ripening and to get better fruit weight and soluble sugar.

Plant Protection

Termites in young plantation can be controlled by application of BHC 10% dust. Rhinoceros beetle and red palm weevil are also attacking as in coconut. The control can be achieved by dusting BHC 10 % in manure pit for former while for later injecting Monocrotophos or a fumigant like celphos will give good control. To control black headed caterpillar (*Nephantis serinopa*) which fees on leaves hiding inside the tunnels in the folds of the leaves, root feeding with 10 mil of monocrotophos mixed with 10 ml of water per tree and releasing the predator *Gonioyis nephantidis* can be done. False smut a disease caused by *Graphida phoenicus* characterized by dark brown or black pustules full of yellow spores particularly under humid conditions can be controlled by spraying 01.% Bavistin or 0.2 % Copper oxyhloride. Fruit rot can be controlled by collected and destruction of infected fruits followed by spraying indofil Z- 78 (0.2%). To prevent birds damage wire gauges can be used.

Harvest, Yield and Processing

Under Indian condition, since the fruit ripening period is not free of rains, the fruits have to be harvested at doka stage during June – August. The fruits harvested at doka stage have to be dipped in boiling water for 5 minutes followed by dehydration in electric oven at 500C for 160 hours or in solar drier to obtain best quality 'Chhuhara' which is the commercial date or produce of commerce. The Chhuhara recovery would be 33 – 35%.

Chapter 16
Subtropical Fruits

MANGOSTEEN - SOIL, CLIMATE, PLANTING, VARIETIES, NUTRIENT AND WATER MANAGEMENT, SPECIAL CULTURAL OPERATIONS, PHYSIOLOGICAL DISORDERS, PESTS AND DISEASES, MANAGEMENT PRACTICES

MANGOSTEEN (*Garcinia mangostana*)

Mangosteen is queen of fruits bearing beautiful at the same time very delicious globose deep purple fruits. Large tree growing upto 10-15 m without angular branches. It is considered as the finest fruit of the world.

- ✰ Native to Malay Arrhipelago and moluccas islands.
- ✰ Fleshy soft, well developed arid is the edible portion.
- ✰ Mangosteen cultivar is popular in Indonesia, Philippines, Burma, Sri Lanka and Malaysia.

Climatic and Soil Requirements

Loves humid tropical conditions *viz.*, high humidity and high temperature without abundant Rainfall and shady environment. Maximum temperature should not go above 35°C. If the temperature goes below 20°C, retardation in growth occur. Severe drought – Sunburn and drying of leaves. Should not have strong wind cause injury to leaves and branching.

Soil : Deep clay to silt loam will be ideal.

Types : Occurrence of natural variability is limited because, the seeds are of asexual origin, they are formed from the nucellar tissue in the 'parthenocarpic' fruits.

Propagation

Sets fruits by parthenocarpy (without pollination and fertilization) and also sets germinable seeds without fertilization propagated through seeds and the plants resemble the mother plant.

When seedlings attain 2 leaf stage – transplanted seedling growth is very slow seedlings do not reach more than 15 cm height ever after 2 years of girth (due to lack of adequate fibrous lateral roots).

Vegetative Propagation

- ☆ Air-layering, grafting or budding
- ☆ Plants are planted at 10 m x 10 m spacing.
- ☆ For young plants, protection from scorching sun should be given.
- ☆ Irrigation needs regular irrigation in places where rainfall is light and good drainage where and rainfall is heavy.
- ☆ Frequency of irrigation – decided on the basis of weather and soil moisture.
- ☆ Mulching without grass and dried leaves – conserves soil moisture.

Harvest and Yield

Harvested at 90 days after full bloom when they are completely mature but still green.They will become soft and slightly purple in clear after 14 days if kept in ambient temperature.If the trees are allowed to sipe on the trees (115 days after full bloom) – they developdeep purple colour). Starts bearing in 7 years.

- ☆ 2 crops (August – October (main crop) and April-June)
- ☆ A 20 years old tree will yield 23-25 kg. Physiological disorders Gamboge and fruit splitting – Physiological disorders
- ☆ Yellow exudation of gum on fruits and branches. Fruit splitting results in swollen arils without mushy pulp.
- ☆ More pronounced in fruits exposed to direct sunlight and in crop that matures in summer. Heavy and continuous rains during fruit ripening favour gambridge and fruit splitting in certain locations.

Chapter 17
Fig

Botanical name : *Ficus carica* L.
Family : Moraceae
Origin : West Asia
Type of fruit : Syconus
Edible Portion : Fleshy
X = 13
2n = 26

SOIL, CLIMATE, PLANTING, VARIETIES, NUTRIENT AND WATER MANAGEMENT, SPECIAL CULTURAL OPERATIONS, PHYSIOLOGICAL DISORDERS, PESTS AND DISEASES, MANAGEMENT PRACTICES

Fig was an important food crop for the ancient civilization of the eastern Mediterranean region. It is a highly nutritious fruit valued as fresh fruit as well as in a dried state. The fruit has a laxative property. The main countries that produce fig are Afghanistan, Greece, lraq, Syria, Spain, Portugal *etc.*,

Climatic and Soil Requirements

The fig is a subtropical fruit the optimum temperature for its good growth being 15.5 to 21°C. Buds of most cultivars requires some winter chilling. In mild tropical and subtropical areas, figs exhibit continuous growth without a distinct rest period. However, the tree usually has a semi- deciduous habit and enters into mild dormancy during October-January. Fruit qualities are also got influenced by climate. A dry climate with temperature especially at the time of fruit development and maturation produces the best quality figs. High temperature like35 to 38oC will result in premature ripening of fruits. Similarly very low temperature will result in splitting and poor quality fruits.

Fig can be grown on a wide range of soil type. But deep, clay-loams are the best suited. Well drained alluvial clay loams or medium black soils are also good for fig cultivation. The fig is one of the most drought tolerant crops. It can tolerate sulphate and chloride salts. Soils having a high lime content produce fruits of better quality suitable for drying. But even a small amount of sodium carbonate in soil is injurious to fig. Major portion of the root system is distributed within 50-60 cm depth and hence the fig can also be grown in shallow soils of 2 feet where other fruit trees cannot be accommodated.

Types and Cultivars

The figs are classified into four types based on the nature of flowers and the methods of pollination.

Common fig

The flowers are pistillate, Fruits develop by parthenocarpy *viz.*, without the stimulation of pollination and fertilization. Kadota, Mission, Adriatic, Brown Turkey, Celeste and Conadria are some cultivars of this type. Poona is one of the most important commercially grown fig. Introduction and evaluation of exotic figs from California at IIHR Bangalore reveals that varieties like 'Deann', 'Conadria' and 'Excel' have uperior fruit and plant characters. These new varieties when grown on 'Brown Turkey' root stock (through chip budding) hold great promise for exploiting marginal lands in arid and semiarid regions.

Capri fig

This type has short styled pistillate and functional staminate flowers. Most caprifigs are not edible, but are grown because they harbour a small wasp *viz.*, Blastophaqa psenes which is necessary for pollination and fruit set in other types like Smyrna fig by transferring the pollen grains from caprifig.

Smyrna fig

It is commercially the most important one. However, the fruits develop only when the flowers are pollinated with pollen from the male flowers of the caprifig transmitted by the Blastophaga wasp. Calimyrna is the common cultivar of this type.

Sanpedro fig

In this type, the first crop is completely parthenocrpic, but the second crop develops only if the flowers are pollinated. The common cultivars of this type are Sanpedro, King and Gentile. In India, common fig is mostly grown. Some of the cultivars grown are Black Itchier, Brown Turkey, Turkish White, Kabul and Marseilles. Yercaud Timla fig is a drought tolerant cultivar. Fruit are large and reddish purple in colour.

Propagation and Plantings

Rooting of hard wood cuttings is the common method of propoagtion in fig. Rooting was the best in cuttings from 3 year old wood with 30-40 cm length and 1.5 cm dia. Cuttings from the base of the shoot and lower part of crown have to be used as they root better. Cuttings are taken during January-February at the time of pruning in North India whereas, the cuttings are taken during rainy season in South India. Fig can also be propagated by air layering, shield or patch budding and side grafting. Focus glomeration rootstock offers resistance to root knot nematode. A spacing of 5-7 M is recommended depending upon the fertility status of the soil for maximum yield. Planting season varies from place to place *viz.*, South India – August – September, Western India – June – July, North India – January – February.

After Cultivation

To keep the trees more productive and to facilitate inter cultivation operations, the fig trees are trained to a desired height and shape. The fig tree bears tow crops in a year, the first crop on the wood of previous season and the second crop on new wood of current season. Pruning is necessary to induce growth of flower bearing wood. The time and amount of pruning are adjusted according to the growth habit and bearing capacity of the tree. Notchin stimulates production of laterals on vigorous upright branches.

Manuring and Irrigation

Fig responds well to manuring. A quantity of 20kg of FYM, 500-600g N and 350-400g P_2O_5 per year per tree can be recommended. Since it is a drought tolerant crop is is mostly grown as rainfed crop. However, irrigation helps to increase the yield. During summer, the crop can be irrigated once in 10-12 days. Frequent irrigation leding to excess soil moisture will cause splitting of fruits. It should be borne in mind that during fruit ripening, the plants should not be given any irrigation because it will result in insipid fruits *viz.*, fruits with bland taste.

Plant Protections

Insect Pests

Leaf feeders : Glyphodes sp

Hypsa ficus

Phycodes radiata

Borers : Olenecamptusbilobus

Thrips : Gigantothrips elegans.

Fig Fly (Lonchaea aristella)

If warranted, these pests can be controlled by spraying endosulfan @2.5ml/lit. which is safer to Blastophaga especially when the caprifigs are interplanted for pollination.

Diseases

Rust

It is caused by Cerotolium fici Small, round brownish to black eruptive lesions occur on the leaves. The rust causes heavy defoliation of leaves. It can be controlled by dusting with sulphur.

Fruitset, Harvest and Storage

It has been observed even in common fig or Adriatic fig, the phenomenon of parthenocarpy is altered by climatic condition of a particular location. Hence there is a possibility of failure of furitset by a particular variety in a particular location.

The parthenocarpic fruitset can be enhanced by spraying 25 ppm of NAA or IBA on the flowers. For Smyrna figs, inter planting of Capri figs should be done for effective fruitset. The fruits should be picked when they are soft and wilt at the neck. If the fruits are picked before proper maturity, milky latex exudes. Fresh figs are highly perishable. Slightly immature fruits are to be harvested for transporting to distant markets. Ripe fruits are picked either form the tree by twisting the necvk at eh stem end or by cutting it or gathered after thy drop. The harvesting season is mid February to June. Yield ranges from 180 to 360 fruits per tree.

Fully ripe fresh figs can be kept only for about a week at 0° C with a 90 per cent relative humidity. To preserve in a dried state, first the fig fruits are soaked in boiling saltwater for half a minute and subjected to sundrying for a few hours. Then they are dried under shade for 8 days and stored in polythene containers. Another from of preservation of fig is drying in an electric drier at 70 – 72° C with prior sulphur fumigation.

Chapter 18
Litchi

Botanical name : *Litchi chinensis* Son.
Family : Sapindaceae
Origin : China
Type of fruit : Nut
Edible Portion : Fleshy aril
X = 14, 15, 16, 17
2n = 30

SOIL, CLIMATE, PLANTING, VARIETIES, NUTRIENT AND WATER MANAGEMENT, SPECIAL CULTURAL OPERATIONS,PHYSIOLOGICAL DISORDERS, PESTS AND DISEASES,MANAGEMENT PRACTICES

LITCHI (OR) LYCHEE

It is a native fruit of warm subtropic zone of China having juice of excellentquality. The fruits are a rich source of sugar (6-20%), protein 0.7% abd faurkt rucg ub fat (0.3%), minerals like calcium (10mg) phosphorus (35mg) and vitamins like vitamin C (64mg), vitamin B1 (0..2mg), B2 (0.03 mg and B3 (0.1 mg) per 100g. The fruits can be canned, made into squash, pickles, and wine or dried to form litchi nut. The tree is a medium large evergreen one with more branches. The fruits are borne in bunches. Each fruit is a one seeded nut, the pericarp is the skin, which has papillate projection like strawberry. The fleshy aril surrounding the seed is the edible portion, which separates easily from seed and skin. China, Taiwan, Thailand, India, South Africa, Australia and Madagascar are the main litchi growing countries.

Climatic and Soil Requirements

A warm subtropical climate is ideal for the growth and flowering in litchi. The winter should be cool as well as dry and devoid of frost (minimum temperature 15°C). This should be followed by fairly long hot (16-22°C) climate with light rainfall during flowering and then by high temperature during fruitset (18-24°) with medium humidity. During harvest the temperature should be around 24-28°C with bright sunlight and high relative humidity. During new flush, the temperature should be

28-30oC with high RH and heavy rainfall. An alluvial soil with good drainage and easy access to the water table is the best for the growth of litchi. The plants have a low ability to transport water from roots to leaves. The soil pH should be around 5.5.

Haak Yip

The fruit is heart shaped each weighing 20-22g, skin is thin, soft and purplish red. Flesh has an excellent aroma and is sweet in taste. It separates easily from flesh. It is the commercial cultivar of Thailand, Taiwan and China.

Taiso

The fruits are eggshaped each weighing 22-26g, bright red skin changing to dull red at maturity, flesh is sweet. It is the commercial cultivar of Australia, Queensland and South Africa (Where it is known as Mauritius).

Waichee

Fruit are small (16-18g) round with deep red skin. It possesses soft flesh with abundant sweet juice. It is cultivated in China and Australia.

Rose secented

Fruits are oblong conical with deep rose pink skin, pulp grayish white. The aril has a delicate rosy flavour. It is an important cultivar of India.

Muzaffarput

The fruits are deep orange to pink with medium juicy sweet pulp. It is grown in India.

Bombai

Fruit is heart shaped, 15-20g in weight usually with another under-developed fruit attached to stalk. Tubercles are carmine red with uranium green interspace. It is an important commercial cultiver of West Bengal.

China

Fruits are globose with a blend of nasturtium red and marigold orange. Fruits are large, 25-27g having a sweet flesh with good juice and pleasant aroma. It is also one of the important cultivars of India.

Propagation and Planting

To avoid the high variability and long juvenile period, seed propagation is not recommended. Semihard and hard wood cuttings of 15-20cm length which were previouslyringed while attached to mother plant and then treated with 10000 ppm of IBA and Ferulic acidwill root easily. Air layering done during spring and early summer season is commercially used.

Stooling, apical grafting, approach grafting, approach grafting and shield budding are other methods of propagation.

The land should be prepared thoroughly by ploughing and leveling. A windbreak should be established around the litchi orchard using eucaluptus, casuarinas, jamun preferably before one year of establishment of litchi orchard.

Pits of 1 M x M 1M x1 M are dug, allowed for a week, filled with 25kg FYM +2 kg of bone meal and 300g of muriate of potash +1 basket of soil from litchi orchard containing mycorrhizal fungi., The spacing recommended is 10 M x 10 M . If the soil is not fertile and the climate is comparatively dry, the spacing can be reduced to 8 M x 8 M. The young plants should be staked.

Irrigation and Manuring

Frequent irrigation is necessary so as to maintain the adequate soil moisture for proper establishment of young plans. If there is moisture stress the plants will not establish. Manurial requirement of lithchi is as follows.

Tree's age (Years)	N (g)	P (g)	K (g/tree)
1	75	25	75
2	100	25	100
3	150	50	150
4	250	75	250
5	250	75	250
6 years & above	600	200	250

For bearing trees also there should not be any moisture stress in soil.

Training and Interculture

Branches with narrow crotches should be removed. Terminal buds in long branching cultivars should be removed to encourage branching. However intensive removal of vegetativegrowth should be avoided as otherwise it will delay the next season flowering. Sorghum stubble, wheat straw, groundnut shells can be used as mulch to conserve soil moisture.

Plant Protection

The bark feeder (Indarbela tertronis) can be controlled by plugging the holes after application of carbondisulphide or formalin. Leaf rollers can be controlled by an insecticide like Nuvan which has fumigant action also. Red rust can be controlled by spraying lime – sulphur 3 times during autumn and 3 times during spring. Leaf galls, caused by eriophid mite Aceria litchi is a common problem. This cab be reduced by pruning and spraying of conventional (or) new avermectin compounds *viz.*, spinosyn, emanectin, spiromesifen *etc.*

Flowering, Harvest and Yield

Litchi trees vegetatively propagated come to flower at the age fo 3 –5 years. The new flush arising in spring end in inflorescence. Fruits are harvested after full maturity and ripening on the tree itself. Fruits are harvested as bunches, precooled and then stored at 0° C to 1°C.

Chapter 19
Avocado

Botanical name : *Persea Americana* Mill.
Family : Lauraceae
Origin : Mexico and West Indies
Type of fruit : Berry
Edible Portion : Pericarp
X = 12
2n = 24

SOIL, CLIMATE, PLANTING, RACES, VARIETIES, FLOWERING BEHAVIOUR, NUTRIENT AND WATER MANAGEMENT, SPECIAL CULTURAL OPERATIONS, PHYSIOLOGICAL DISORDERS, PESTS AND DISEASES, MANAGEMENT PRACTICES

Avocado, one of the most nutritive fruits, has orighinated in Central America. Avocado is now grown in most of the tropical and subtropical countries. The pulp of Avocado the so called butter fruit, has a buttery consistency and the fruit has very high fat content (26.4 g/100g). As the sugar content is low, it can be recommended as a high energy food for diabetics.

Climatic and Soil Requirements

Avocado trees of the West Indian race perform well in humid tropical climate, but the other two races *viz*., Mexican and Guatemalan fail to flower or set fruit in the tropics. On the other hand, the West Indian race sets little or no fruit in subntropical climte. In between Mexican and Guatemalan races, the Mexican race alone survives in regions where minimum winter temperatures goes to – 0.5°C to 3.5°C. If proper race and cultivars are chosen, avocadoescan thrive and produce a good crop in climatic conditions ranging from true tropical to warmer parts of the temperate zone.

Though avocadoes can be grown successfully in varying soil conditions, they are extremely sensitive to poor drainage and saline conditions. They will be happy in soils with pHranging from 5-7.

Mexican Race

This group is characterized by small fruits weighing less than 250 g and ripening in 6-8 months after flowering. Oil content of the fruit is 30 per cent, the highest of all the three races.

Guatemalan Race

Fruits are fairly large each weighing upto 600g and ripen in 9-12 months after flowering.

Oil content ranges from 8 – 15%.

West Indian Race

The fruits are medium sized and ripen in 9 months after flowering. The oil content of fruit is low, ranging from 3 – 10%.

Cultivars

Fuerte

It is the most popular cultivar of avocado. It is a hybrid of the Mexican and Guatemalan races. Fruits are pyrifrom, each weighing between 225 and 450 g having 18 to 26 per cent oil. It is fairly resistant to cold, better suited to subtropics than tropical climate. It belongs to group-B.

Hass

It originated as a seedling from the Guatemalan race. It matures much earlier than Fuerte. Fruits are medium sized, roundish and turn purple on ripening. This is also more suitable to subtropical climate. It belongs to group –A.

Pollock

A West Indian race bears large fruit that weights upto 1 kg or more having an oil content of 3-5 per cent suited to be grown in a tropical climate.

Purple

It belongs to the West Indian race. Fruits are pear-shaped, skin is deep crimson ormaroon in colour. Suited to humid tropics.

Green

This belongs to the Guatemalan race. Fruits are large, oval, with yellowish green skin. It is suited to subtropical climate.

TKD 1

Developed at Horticultural Research Station, Thadiyankudisai of TNAU, Tamilnadu.

The fruits are medium sized and round. Trees upright and semispreading hence suited for high density planting. Yield 264kg/tree. Fruits are sweet TSS8o brix, fat 23.8%, protein 1.35%.

Propagation and Planting

Avocado is normally propagated by seeds. The viability period of avocado seed is very short (2-3 weeks), and can be improved by storing the seed in peat or sand at 5° C. Removal of the seed coat before sowing helps to speed up the germination. The seed can also be split lengthwise into 4-6 parts, leaving a piece of embryo on each. Avocado can also be propagated vegetatively by cutting and grafting. The Mexican race is relatively easy rooting whereas, the West Indian race is relatively easy rooting whereas, the West Indian race is quite difficult to root. The Guatemalan race is intermediate in rooting ability of cuttings. Cleft, whip and tongue and whip grafting are the most successful methods. The normal planting distance for avocado is 6-12 M depending on the vigour of the cultivar.

Manuring

Avocado requires heavy fertilization. Application of nitrogen is the most essential. Nitrogen deficiency causes restricted growth, with reduced, pale coloured leaves and smaller fruits. An average crop of avocado removes 40 kg N, 25kg P_2O_5, 60 kg K_2O, 11.2 kg Ca O and 9.2 kg MgO/ha from the soil. Therefore, to maintain the soil fertility for getting consistent yields, it becomes necessary to replenish these nutrients.

Problems in Fruitset

Avocado starts bearing at 5-6 years after planting and has a marked tendency to biennial bearing which is prevalent in a number of other fruit trees. But there is specific problem in fruitset as far as avocado concerned.

In avocado, the inflorescence is a compound panicle. The individual flowers are morphologically bisexual having fertile male and female organs. But they exhibit dichogamy *viz.*, the male and female organs coming to maturity at different time thereby avoiding selfpollination of an individual flower. In dichogamy, they are protogynous *viz.*, the female parts coming to maturity before male organs. The type of dichogamy in avocado is a complicated one unique to avocado-the diurnally synchronous dichogamy. The female parts of all flowers that open at a time in a particular tree will mature simultaneously and hence behave functionally as female flower. The male parts of same flowers will come to maturity when the flowers open next time and hence all of them behave as male flowers during that period. By this the cross pollination between flowers of the same tree are also ruled out. The situation is further worsened by the fact that all the trees of a particular group will be exhibiting the same sex phase at a particular time and the opposite sex phase during the next opening of the same flower. So if the trees of single group are planting in mass, they will not set fruit and each group requires inter planting of trees of mother group, the two groups being compatible with one another.

Based on this unique flower behaviour, avocado cultivars can be divided into two groups A & B. In group A, first opening takes place in the morning, second opening during the afternoon of the following day. In group B, first they open in the afternoon then again next morning. Therefore, every morning A-pistils can be fertilized by B-pollen, while during afternoon B-pistils are ready to receive A-POLLEN. Honey bees are the chief pollinating agents.

Interculture and Weed Management

Deep cultivation in avocado orchards should be avoided because of surface roots. Intercropping with legumes or shallow-rooted crops can be done in young orchards which can smoother weeds also.

Avocado trees are pruned sparingly mainly by heading back the central shoot in upright growing cultivars such as Pollock, to develop a spreading habit. Branches are thinned and shortened in spreading cultivars like Fuerte.

Plant Protections

Pests

Mites, mealy bugs, scales are the important insect pests of avocado. Spraying of systemic insecticides will effectively check these pests.

Fruitspot

It is caused by Colletotrichum gloeosprotioides. Infection results in shedding of young fruits. Ramaining fruits become deformed. This can be controlled by spraying of Indofil- M.45@2g/lit. Controlled atmospheric storage of fruits in 2% O_2 at 7.2°C for 3-4 weeks will prevent the development of the fungus in storage.

Root rot

Root of avocado is caused by Phytophthora cinnamomi. Soil drenching of Ridomil (1gai/10 lit) controls root rot.

Harvest and Storage

Fruits should be cut from the tree using hand clipper or a long pole equipped with a clipper and cloth catching bag. Average yield is about 100-500 fruits per tree. Avocado fruit does not soften while on the tree, but does only after it is picked. The matured avocado fruit ripen in 6-12 days at 20°C.

Chapter 20
Durian & Carambola

SOIL, CLIMATE, PLANTING, VARIETIES, NUTRIENT AND WATER MANAGEMENT, SPECIAL CULTURAL OPERATIONS, PHYSIOLOGICAL DISORDERS, PESTS AND DISEASES, MANAGEMENT PRACTICES DURIAN

[*Durio zibethinus* Bombacaceae]

Native to Borneo, an Island in the Malaysian region. It is tall evergreen tree of humid tropics with ovoid oblong fruits, each weighing 2-5 kg characterized by trough fibrous rindcovered with greenish, coarse, sharp, pointed, short spines. The edible portion inside the fruit is buttery with yellowish while pulp with very strong odour. People have strong like or dislike for this fruit. Many like it, because if its sweet, delicious and filling taste, while others dislike it for its highly objectionable odour. The penetrating odour is comparable to that of rotten onion and is unacceptable to many. The chemicals that are responsible for the flavour are hydrogen sulphide, ethyl hydrosulphide and dialkyl polysulphide.

100 g of edible portion

28.8 g CHO

2.5 g protein

2.5 g fat

20 mg Ca

63 mg P

37 mg Vitamin C

Indonesia, Philippines, India, Thailand, Myanmar and South China.

The fruits are believed to have aphrodosiacal qualities. Root decoction is used to cure fever and leaves for curing jaundice.

Cultivars

Frog, Gibbon, Longstem, Deception. Cha-Nee, Golden Pillow are important varieties of durian.

Climate and Soil Requirements

Perform well in areas with high humidity and high temperature, frequent showers and low wind velocity Annual Rainfall 200 cm, but heavy rains prior to flower initiation affect normal flowering. The trees cannot tolerate temperature below 8°C. Deep fertile soil or clay loam heavy good drainage and organic matter are the best suited. Salt in soil or water will result in fruit drop.

Propagation and Planting

Vegetative propagation by patch budding on seedling rootstock is recommended. Seeds should be sown immediately after extraction from fruit as they loose viability very rapidly. Though propagation through cuttings is easy, not recommended as the adventitious root system will be insufficient to bear such tall stem. Plants of atleast 2 cultivars have to be planted for effect pollination. Spacing is 10-12 m.

Pruning

The tree usually takes pyramidal shape by itself. However, after harvesting the tree can be cleared off dried and old twigs.

Manuring

It needs manuring at early stages of crop growth. Upto 7 years absorption of nutrients and retention on the vegetative part is more – 20 kg Nitrogen, 5 kg Phosphorus and 10 kg K/ha Potassium can be applied during 3rd year and can be increased to 40 kg, 10 kg and 20 kg of NPK during 7th year. Beyond 7 years. Only very little quantity can be applied.

Plant Protection

Hawk moth is serious problem, larvae feed on leaves, soil incorporation of Lindane 1.3 D and creation of light traps recommendation, harvest, yield and post harvest management. Grafts produce crop in 4-5 years, seedlings take 10 years. Fruiting available 2 times/year. Durian has cauliflorous flowering in older branches. Self incompatibility noticed in Durian.

Yield – 100 fruits/year. Harvesting done when the fruits are smooth, flat with far – apart spines.

Ripe fruits last for 4 days under refrigeration, while mature fruits can be stored upto 14 days at 10-15°C.

CARAMBOLA

Botanical name : *Averrhoea carambola* L.
Family : Oxalidaceae
Origin : India-China
Type of fruit : Berry
X = 12
2n = 24

Carambola is a warm tropical and subtropical fruit with five star shaped projections and having juice with characteristic acidic taste. It is also known as five corner fruit or five edge fruit. The juice content is about 60 – 70%. The acidity is mainly due to oxalic acid and malic acid. One hundred gram of fruit contains 3.5 – 11.0 g sugar, 0.75 g protein, 560 IU of vitamin A.

The probable origin of carambols is Malaysia or Indonesia. Besides consumed as dessert fruit, carambola can also be served as juice, jam, jelly and pickle. The star shaped cross section slices of fruit can be used in fruit-salad and to decorate cakes. A close relative of carambola is Bilimbi (Averrhoa bilimbi) which produces more acidic fruits.

Climatic and Soil Requirements

Carambola loves a warm, humid condition with long hours of sunshine. Though it can be grown in a wider range of soils *viz.*, right from sand to heavy clay loam, it will be at its best on well drained deep clay loam with a pH of 5.2 to 6.2.

Cultivars

Golden Star

Developed by Florida, Queensland, Singapore, Taiwan and Japan. The wing edges are slightly rounded. Fruits are golden yellow, crisp each weighting 90m-200 g.

B1

Developed at Malaysia having lemon yellow fruit which are crisp each weighing 100-300 g. The edges are rounded.

B6

Another variety developed at Malaysia with orange fruits. The edges are slightly rounded.

Maha

Commonly grown in Florida. Fruits pale lemon yellow each weighing 100-200g. Besides this a number of varieties like Thai kinight, wheeler F wang Tung are grown in **Florida.**

In Australia a number of local selections like Jungle Gold. Chujuba, Giant Siam are being cultivated.

Propagation and Planting

Approach grafting, wedge grafting and airlayering of one year old terminal shoots (treated with IBNA 10,000 ppm at the time of layering) are the common methods of vegetative propagation recommended. The plants can be planted in pits of size 1 M x 1 M x 1 M dug at a spacing of 8 M x 6 M.

Irrigation, Pruning and Manuring

Caramnola is very sensitive to water stress. It demands very high quantity of water (2000 lit/week/tree/in maximum demanding months). Mulching the tree basin will jelp to prevent moisture stress. In young trees 4-6 wide – crotch angled branches are allowed. Low hanging criss – cross branches should be pruned one year after planting. Each tree is applied 40 g N, 10g P and 70 g K in 4 equal splits during April, June, September and January. Every year it can be increased in the same ratio to reach 600 g N, 120 g P and 1000 g K/tree at 8th year.

Plant Protections

Fruit borer can be controlled by spraying chlorpyriphos 1 ml/lit. To control fruit sucking moth, enclosing the fruits with mesh less than 10 mm is recommended. Cercospora leaf spot can be controlled by Indofil M. 45@ 2g/lit.

Flowering, Harvest and Yield

In India flowering is seen in three flushes April – May, July – August and September – October and fruits are harvested in July – September, November – December and January – February. Grafted trees start bearing 3 year after planting. The yield ranges from 200 – 300 kg/tree from 7-8 years onwards. Even at 21°C, carambola fruits can be stored for two weeks.

Chapter 21

Rambutan & Longan

Botanical name	:	*Nephelium lappaceum* L.
Family	:	Sapindaceae
Origin	:	Malaya
Type of fruit	:	Berry
Edible Portion	:	Aril

X = 11

2n = 22

SOIL, CLIMATE, PLANTING, VARIETIES, NUTRIENT AND WATER MANAGEMENT, SPECIAL CULTURAL OPERATIONS, PHYSIOLOGICAL DISORDERS, PESTS AND DISEASES, MANAGEMENT PRACTICES

Rambutan is called "hairy litchi" as the fruits have red soft spines on the surface. Thebwhite translucent, sub arid – seet flavoured aril is the edible flesh of the fruit. Its appearance differs from the litchi by having red and soft spines (hairs) covering the whole surface of the fruit.

Origin

Malay archipelago, from there it has spread to South East Asia, Central America and Africa. Rambutan growing countries are Thailand, Myammar, Sri Lanka, India.

100 g fruit contains 14.5 CHO

0.9 g protein

0.1 g fat

31 mg/100 g Vitamin C.

TSS : 14.9 – 21.2%

Uses

Rambutan mainly serves as fresh fruit or a dessert mixed with other fruits or used for canning or made in to fruit syrup. The oil of seed is used for soap making. The root has medicinal uses while the tannin it contains is used for dyes. The whole tree with its beautiful leaves, flowers and colourful fruits also serves as an environmental tree in landscape. Tree is large, round-topped, much branched every green tree reaching upto 15 m or more in height. Fruits are one-seeded, large, ovoid in shape, 10-12 fruits are borne in clusters. Fruits are 3.5 – 8 cm long, 2-3.5 cm wide, soft, with curved spines weighting 14.3-43.7 g.

The soft curved spines (hairs) covering allt he pericarp may be red or yellow in colour. Inside the fruit, a big seed is surrounded by pearl-while aril/flush) which is fixing, subacid-sweet in flavour but the taste is rather flat or insipid as compared with the sweeter litchi.

Climate and Soil

It is a high-temperature bring plant. A warm humid tropical climate would be ideal. Mean minimum temperature 23°F. Mini temperature should not be below. 13-15°C, it comes well under humid condition. It requires 2000-2800 mm annual rainfall which should be evenly distributed over the year.

A rich sandy or clay loam with high organic matter and good drainage will be ideal for the growth and flowering. When reference is insufficient, the land should be irrigated in time to provide sufficient moisture to the soil.

Cultivars

Rohug-Rian – Thailand cv.

- ✰ Oval bright red fruits spines are red which green tip. The aril is juicy, tough and sweet.

Seematjan

- ✰ Fruits deep red with soft curved spines.

BR.1 (Boting Rambutan No.7)

- ✰ Fruits are big (30g), ovate, red *etc*. spines are fine and dense. Aril is juicy, crisp and sweet.
- ✰ New cv. Developed in China.

Propagation

Seed

Dioecious tree, about 1/3rd of the total number of the progeny become male plants. Seed propagation not recommended as seedlings are very variables and may produce low quality infirm fruits. Used only for raising seedling rootstock for grafting seeds are sown immediately when taken out from fleshy fruits because they loose viability within a few days, if left day.

Vegetative Propagation

- ✰ Patch or forkert budding on 1-2 year old seeding rootstock is recommended.
- ✰ Air-layering is another method.

Planting

Spacing given for planting 10 m x 7 m Planting of different cultivars is advised to encourage pollination. Temporary shading of young plants necessary.

Training and Pruning

Modified leader system of training is practiced. Light unusual pruning of crossing branches, dead and diseased twigs and water shoots are suggested.

Manuring and Interculture

Fertilizer application done similar to litchi. Mulching can be done to conserve soil moisture. Given manuring cover crops may be grown and twened into the soil to supply organic matter.

Flowering

Rambutan is androdioecious with separate male and hermaphrodite frees. The male is seldom found since now a days rambutan is planted from budded material. Hermaphrolite trees are often referred as female because it bears fruits. Flowers are apetalons, greenish white in colour, about 2 m in diameter.

Male Flowers

5-7 stamens, each stamen has a whitish to mentose filament with a yellowish bilobed anther. At the centre of the file is the abortive ovary which is highly pubescent.

Hermaphrodite Flower

6-7 stamens, but the anthers do not dehisce to release the well developed pollen grains inside. Functionally it serves as female flowers. Ovary is two or three-lobed and bears a bifed or trifid stigma.

Flowers at all the time of the day, but peaks at 6.30 am. Fruits can be harvested 13 weeks after fruitset. Some rambutan trees may set a few parthenocarpic fruits which are smaller and somewhat flattened in form but the aril maintains normal flavour as the one-seeded fruits.

Harvest and Yield

Average yield 25-35 kg/tree/year fruit colour is often used as guide for harvesting.

Storage

- ☆ Non-climateric fruit. Fruits can be kept for 4 days at 25-30°C.
- ☆ Desiccation, loss of red colour, browning and drying of spintern fleshy pliable spines) are the principal causes of deterioration.
- ☆ Storage of fruit in sealed polyethylene bags at 12°C retarded skin colour loss and extended shelf life (18 days).

Plant Protection

- Litchi stink bug, bark borer, fruit weevil, leaf roller are major pests.
- Downy mildew, anthracnose are the diseases.
- Both can be controlled by application of pesticides and fungicides whenever necessary.

LONGAN

Longan is a close relative of lychee having similar growth and fruiting habit, but somewhat hardier than lychee. The fruits are smaller, smooth surfaced with yellow tan to brown colour. It originated in subtropical region of China or in between Myanmar and India. Fruits contain 25.2% carbohydrate, 1% protein, 0.5% fat, vitamin A (28 1U/100G) and vitamin C.

Climatic and Soil Requirements

Longan requires a warm subtropical climate with high rainfall. The winter should be short and frost free. The summer should be long, hot, humid and wet. The best temperature for flowering and fruitset is 20 to 25%C. Temperature above 40oC will cause damage to fruit resulting in fruit drop. A deep well drained fertile soil with pH 5.5 to 6.0 with low salinity will be ideal for its growth.

Cultivars

Fu yan (Lucky Eye)

It is the major cultivar of fujian province of China. Fruits are large (18g), thin skin, crisp flesh and small seed. It is the best variety for canning.

Wu long line (Black Dragon peak)

Medium sized fruits (15g) with thick skin, good flesh having sweet flavour.

Wu yuan (Black Round)

Fruits are medium in size (15g), seed large, flesh soft and juicy.

Daw

It is the popular cultivar of Thailand. Fruit is large with big seed, thin skin and crisp sweet flesh.

Propagation and Planting

Longan cultivars have to be propagated by air layering or marcotting the most recently flushed growth with maximum maturity at about 45-60 cm from the tip. Land preparation has to be done 3 to 6 months ahead of planting. Water ways (grassed) to carry run off water, and formation of terraces especially in slopes and mounds of 0.5 M to help drainage have to be formed. A planting distance of 6 M x 6 M can be allowed in hdp system. Plants should be planted in pits of 1 M x 1 M x 1 M size.

Training, Pruning, Irrigation and Manuring

Early growth is to be encouraged so as to get a good tree size and canopy within 3-4 years. Frequent irrigation and regular application of nutrients will help this. Organic manures like poultry manure should be supplemented with fertilizers like Urea. The trees can be trained to height of 2M and 3-4 vigorous young growth with wide crothes are allowed to form main frame work. Each lateral can be allowed to have w secondaries. The trees should be trained to get food light penetration through the canopy. Each tree should be earthed up to make mound (a height of 0.8 – 1.0M). Irrigation is given to encourage bud burst. Before flower emergence it should be withheld. Again irrigation is to be continued from panicle emergence, flowering, fruit set, harvest and post harvest flushing. Aagin when the tree undergoes a rest period, irrigation ahuld be stopped. Just like irrigation, the fetilizer appliation should also be done 3 –4 times during the different growth periods in a year. For a five year old tree 625gN, 150g P and 800 g K have to be applied and increased by 20% every year till 10th year.

Interculture

Alternate bearing is a common phenomenon. In an „on" year, 40% of the flower spikes and 30% of young fruits can be thinned. In an „off" year, a portion of th eshoots produced is spring should be thinned to reduce the competition between shoots and fruits.

Plant Protections

Fruitfyl (Bactrocera dorsalis) is a serious pest, which can be controlled by any safe insecticide like malathion. (Refer Mango section for details).

Harvest and Yield

Longan fruits are harvested after full maturity and ripening. A fully matured tree yields 100 – 120 kg/year. The fruits can be stored for 4-6 weeks at 5°C.

Chapter 22
Apple

Botanical name : *Malus x domestica* **Borkh.**
Family : Rosaceae
Origin : Asia Minor to Western Himalayas
Type of fruit : Pome
Edible Portion : Fleshy thalamus
X = 17
2n = 34, 51

SOIL, CLIMATE, PLANTING, HIGH DENSITY PLANTING, VARIETIES, NUTRIENT AND WATER MANAGEMENT, SPECIAL CULTURAL OPERATIONS, PHYSIOLOGICAL DISORDERS, PESTS AND DISEASES, MANAGEMENT PRACTICES

'An apple a day keeps the doctor away'. This is the old proverb which significe the nutritive importance of apple in human diet. Apple is a rich source of easily assimilable carbohydrate (13.4%) and it is also fairly rich in calcium (10mg/100g), phosphorus (14 mg/100g), and potassium (120mg/100g). it supplies vitamin B and C. apple has been under cultivation since time immemorial and today more than 80% of the world's supply is produced in Europe where the major producers are Italy, France, and Germany. Other countries which also produce apple are Hungary, USA, N.C.America, China, Old USSR, India, Spain, Switzerland, Iran and South America.

Climatic and Soil Requirements

Apple is essentially a temperate crop and grown in temperate region of the world. Under subtropical zone at an altitude of 1600 – 2500 M above MSL also, apple can be grown. Very low temperature during the bud rest favours better crop production. The typical temperate types require 1000 hours of uninterrupted chilling below 7° C for winter rest. After the bud break, during the growth, long day hours with high light intensity, warm days (not hot days) *viz.*, 12 – 15° C and cool nights (not freezing nights) *viz.*, 7 – 8° C are favourable for production of quality fruits in large quantities. The freezing soil temperature can kill the trees. A well drained, slightly acidic (pH 6.5 – 6.7) loamy soil with good depth (45cm or more) is considered to be ideal for apple culture.

Propagation

Apple varieties are propagated by whip and tongue method of grafting. The root-stocks are either related species such as *Malus sylvestris* (crap apple), *M. prunifolia, M. sikkimensis* or their hybrid derivatives or seedling progenies of cultivated varieties. The main aims indeveloping rootstocks are either to dwarf the trees, or to have resistance to below freezing soil temperature or high soil temperature, to wooly aphid, root knot nematodes *etc*. The East Malling Research Station in England developed Malling Stocks (M). John Innes Horticultural Institute, Jointly Merton and East Malling Research Station jointly developed rootstocks (MM) which, are clonally propagated by either cutting or stooling. Some of the important rootstocks developed for specific purposes are as follows:

- ☆ M9, M – 27 : Dwarfing
- ☆ M7, MM-106 : Tolerant to below freezing (-40° C to – 35° C) resistant to wooly aphids.
- ☆ Northern spy : Resistant to wooly aphids (*Eriosoma lanigera*).
- ☆ Robusta-5 : Developed at Ottawa. Canada as a selection of *Malus robusta* – Resistant to wooly aphid. Other rootstocks used are Merton 779, 793.

Varieties

In India at himachal Pradesh, varieties like king of Pippins, McIntosh, Golden Delicious, Red Gold, Starking Delicious, Yellow Newton and Granny Smith are grown. In Jammu and Kashmir, Irish Peach, Cox Orange Pippin, Kerry Pippin, Ambri, Lal Cider, Golden Delicious, Lal Ambri, Red Delicious, Sunehari and Razakwar are important varieties under cultivation. In utter Pradesh and Fanny, Cortland, Early Shanburry, Golden Delicious, McIntosh, Red Delicious and Buckingham are important varieties. In the hills of tropical region where warm winter conditions exist, Parlin's Bearty and Tropical Beauty have been judged as the best performers. They require only less chilling for bud break and flowering.

Red Delicious

It is a midseason variety. Widely grown cultivar throughout the world. Evolved as a chance seedling in Iowa, USA. Fruits are large, oblong conical with 5knob like projections at calyx end. Smooth skin covered with red streaks on a plain background. Flesh creamy white, tender, crisp, fine grained, sweet and highly aromatic. A number of strains like Starking Delicious (Royal Delicious), Ruby Red, Red Spur, Red Chief, Richard Delicious have been developed from this cultivar.

Jonathan

It is also a mid season variety, fruits are round to conical in shape medium to small in size and reddish in colour.

Golden Delicious

It is a late variety with moderate vigour. Fruit medium to large, oblong, skin golden yellow with russetted prominent small dots scattered all over. Flesh creamy white, firm, crisp, sweet with a blend of acidity. It is a good pollinizer for all Delicious group.

McIntosh

It is a mid season variety. Trees are vigorous. Fruits are medium in size, oblate round, skin smooth, shining carmine colour on pale green background. Flesh White, develops red shade after storage, tender, crisp, sweet with a good acid blend, juicy and mild flavoured. But this cultivar is scab susceptible.

Bramely Seedling

It is an English cooking apple ideal for stewing especially baking. Skin deep green, waxy sometimes with a slight orange red blush.

Jona Gold

(Triploid) Golden Delicious x Jonathan. Developed in New York. Fruits large with attractive red strips and have good storage life.

New Jona Gold

A virus free clone of Jona Gold at Japan with super grade having high red colour intensity.

Kent

A hybrid between (Diplod) Cox and Jonathan. Late maturing variety; fruit slightly conical with dark red colour on a green background. For this, Golden Delicious and Sparten are good pollinizers.

Chaubattia Anupam

Developed at Chaubattia from a cross Early Shanburry x Red Delicious. Fruits medium in size, skin smooth, thin shining yellow almost entirely flushed are striped red; flesh is firm, crisp, creamy white, juicy, sweet with distinctive aroma.

Shamrock

A cross between irradiated McIndosh x Starkspur strain of Golden Delicious. It is green apple identical in appearance and taste with Granny Smith but matures six weeks earlier.

Romus-2

Released from Romania, resistant to scab. Early maturing (mid july onwards). Fruits are medium sized (120g) red in colour with good flavour.

Kodaikanal-1

A clonal selection made from Parlin's Beauty at Horticultural Research Station, Kodaikanal in Tamil Nadu. Adapted very well to warm winter conditions prevailing in South Indian Hills. Fruits is medium weighing 150g, global in shape with crisp flesh which is moderately juice. Fully rips fruits are attractive yellow in colour with crimson drop.

Applethrope Summerdel

Cross between Delicious x Earliblaze, fruits medium in size 6.5 – 7.5 cm dia. Globose to flat globose. Flesh creamy white to yellow, medium coarse texture, firm, juicy and sweet.

Developed at Granite Belt Horticultural Research Station, Queensland, Australia.

Richelien Apple

Developed at Canada Agricultural Research Station by a multiple cross involving Melba, McIntosh, Jonathan, Rome Beauty and *Malus floribunda*. Even after full ripening, it hangs on the tree for 10 days without loosing its firmness. It has good flavour.

Freedom

Developed at New York. Highly resistant to scab. Fruit is large in size and oblate in shape with firm moderately coarse flesh.

Planting

Pits of size 60 x 60 x 60 cm are dug at a spacing of 5 M x 5M. For effective fruit set (since self incompatibility exists in most cultivars) pollinating varieties are interplanted. Usually one in six or one in nine plants is used as pollenizer in self-unfruitful varieties. In intensive system, planting is done at 4M X 2.5 M or less. A complete row of pollenizer is used for every two rows of main variety.

High Density Planting (HDP) and Medium Density Planting (MDP)

For this dwarfing rootstocks like M9, M4, MM106 and M26 have to be used. The grafts are planted at a close spacing of 2M X 1M, so that 5000 plants/ha. Are accommodated. But incomparison with M. D. P. where in the total number of plants accommodated was 1666 trees/ha (3 M X 2 M), the H. D. P. did not give much yield. An additional yield of 50% more than HDP can be obtained from MDP. The fruit colour and quality also better in MDP than HDP.

Training and Pruning

Training is done to give a desired shape as well as to get a canopy for intercepting maximum sunlight so that all the portions of the tree get sunlight and become productive. The trees are trained as central leader, open centre, modified leader, cordon, dwarf pyramid, espalier, tatura trellis and Lincoln canopy. The apple bear fruit buds in a past season growth called spur. In the established spur system, permanent fruit spurs are obtained on laterals by heading back the leader every year. The strong erect laterals near the central leader are removed leaving wide angled vigorous laterals for formation of spurs. In regulated syatems, the crowed weak and shaded branches are removed leaving leader and strong laterals are grow.

Manures and Manuring

Apple trees require all the mineral elements. Depending on the soil type, rootstock, a quantity of 500g of N, 250g of P, 750g of K/tree can be applied. Deficiency of calcium will lead to physiological disorder in apple fruit called 'bitter pit' or 'cork spot'. Bitter pits appear as slight indentation in the skin usually towards the calyx of the fruit. These areas turn brown and soft dried pits of collapsed tissues develop. In fruits, colour changes and cracks develop. Spraying 0.5 % calcium chloride during the later development of the fruit. 4 times at 14 days interval will help to reduce the deficiency symptom. Boron deficiency also causes internal browning of fruit and corking spots on the flesh (internal cork) and sometimes as cork tissue in the skin (external cork). Sparying sodium borate @ 10g/lit. 3 times during flowering and fruit set will help to correct the disorder.

Special Cultural Operations

The drip or trickle irrigation system has high water use efficiency in apples. It also suits the tree in hill slopes. Mulching can be done with straw, hay or sawdust and other organic matter. Weed control can be achieved by application or herbicide such as Pitezin (Atrazine + Simazine + Propazine) @ 8 – 10 kg/ha, after weed emergence in 1.5 M band along tree rows. Nearer to rootzone littering of straw is advised to maintain organic matter and earth warm activity. Since herbicide can result in a significant reduction of earthworm population.

When there is chilling deficiency , the bud break will be poor. For bud break, application of Sandolin – A (Dinitro Ortho cresol) at 0.5% in 2% mineral oil emulsion during January is recommended. After bud break, flowring and petal fall. Alar @ 3000 ppm is applies which will encourage the duration of flowering.

To prevent fruit drop, AVG (Amino ethoxy Vinyl Glycine) 200 pm is sprayed which can increase the fruit set to 10 fold when applies between pink and petal fall.

Plant Protection

Pests

1. Aphids (*Eriosoma lanigerum* and *Aphis pomi*)

Symptoms

They feed on new wood and roots to farm galls. Cold and wet condition favour the development of this pest.

Management

- ☆ Destruction of eggs before hatching should be done by spraying tar oil or DNOC – petroleum oil, thoroughly wetting the bark of the tree. Spraying of organophosphorous insecticides (malathion 0.2%, phosphomidon 0.2% or dimethoate 0.2%) at green cluster stage will be effective.
- ☆ Use wooly aphid resistant rootstocks like Northern spy, Robusta - 5
- ☆ Its main natural enemy, are the parastoid, *Aphelinus mali* is an important control agent.

2. Blossom Thrips (*Thrips flavus*)

Symptoms

The attacked flowers show withering symptoms, as a result either the fruits do not set or may fall off in the early stage of development. Heavily infested bloom produces distorted flowers that open on one side. Excreta are often deposited near the feeding site, which provides a suitable site for fungal infection.

Management

- ✰ *Chrysopa sp*. And lady-bird beetle (*Coccinella septumpunctata*) are the most voracious predators of thrips.
- ✰ Pre bloom spray at green tip stage with fenetrothion (-0.05%) reduces the population of thrips.

3. Red Spider Mites (*Panoychus ulmi*)

Symptoms

The mites suck sap from the leaves, which may turn bronze. In severe infestation. The growth of the plant is markedly reduced.

Management

The component of integrated management of red spider mites is the regulation of predators *Typhlodromus pyri* and *Amblyseius andersoni*

Diseases

1. Apple Scab (*Venturia ineequalis*)

Symptoms

Velvety brown to olive green powdery lesions which turn mousy black with age, appear on the leaves and lead to premature yellowing of leaves, defoliation and fruit drop. Scab spots n fruit results in misshapen and knotty fruits. Fissures or cracks develop in the scabbed areas which allow the entry of other pathogens, causing rot of fruit.

2. Powdery Mildew (*Podossphaera leucotricha*)

Symptoms

Whitish powdery growth develops on both sides of leaves and twigs. The affected leaves are desorted in shape and small in size; become hard and brittle. The powdery coating on twigs finally disappears and a brown. Felt – like covering with black fruiting bodies is seen, in case of severe infection, leaf fall and premature fruit drop may occur. The young infected fruits show signs of rusting.

Management

Destruction of over wintering fungus structures reduction of inoculam by pruning shoots. Spraying triadimephon (0.05%) or pyrazophos (0.021%).

Harvest and Yield

Fully mature when start developing colour can be harvested. The normal yield ranges from 100 – 150 tonnes/ha in medium density planting.

Chapter 23
Pear

Botanical name : *Pyrus communis*
Family : Rosaceae
Origin : China
Type of fruit : Drupe
Edible Portion : Stalk of fruit and thalamus
X = 8
2n = 16

SOIL, CLIMATE, PLANTING, VARIETIES, NUTRIENT AND WATER MANAGEMENT, SPECIAL CULTURAL OPERATIONS, PHYSIOLOGICAL DISORDERS, PESTS AND DISEASES, MANAGEMENT PRACTICES

The probable origin of the Pyrus genus is mountainous China. From there it moved to East and West and in the different ecological conditions, the evolution would have taken place. Pear is next to apple in the temperate fruits. It is a rich source of carbohydrate as sugars, starch and cellulose and minerals like calcium (8 mg/100g) phosphorous (15mg/100g) and sulphur (14mg/100g).

Climatic and Soil Requirements

Pear is adapted to a wide range to climatic condition from as low as –26° C (in dormancy) and as high as 45° C (in growing period). However most of the pear cultivars require about 1200 hours below 7°C during winter to complete their chilling requirement in order to flower and fruit to a satisfactory level. However after bud break, at blossoming and fruiting phase, below freezing temperature will affect the crop severely. Because of its tolerance to wide range of climate and soil, it is grown both in temperate and subtropical conditions. In the tropical beltpear can be grown at an elevation of 1300-2100 M above sea level. Deep, well drained, fertile, medium textured clayey loam soil is the bet for pears. A neutral pH range of 6.0-7.5 will be ideal. A minimum soil depth of 180 cm is required. When compared to apple, pear is less tolerant to drought but more tolerant to wet soils.

Varieties

Pear cultivars can be classified in to 3 groups *viz.*, European types, Asian types and the hybrids of European and Asian types.

The Asian types (oriental pears) are mainly derived from *Pyrus pyrifolia, Pyrus ussuriensis* or their hybrids. The European types are derived from *Pyrus communis.*

European Types (Tail Pears)

Bartlett (or) William's Pear

It is the most popular commercial cultivar throughout the world (except Chin & Japan). It requires more chilling hours (1500 hrs) for bud break and flowering. The fruit is ovate pyriform, medium large, green at the time of harvest turning to bright yellow after ripening. Flesh is white, melting, firm and juicy. It has originated in England. High yielding clones like Bartlett Improved I, Verona-25, Ferrera have been identified.

Anjou

Originated in France. It is fairly resistant to very low temperature and fire blight. Large fruit, skin bright green when harvest and turning to greenish yellow on ripening. Flesh is fine, mildly acidic. Fruits have high dessert quality and very long keeping quality.

Flemish Beauty

The trees are bigger with more branches. Fruit is large obovate and smooth. The flesh is pure white, very juicy free of grit cells. It is a self-fruitful variety. It can also be used a good pollenizing cultivar.

Max Red Bartlett

A bud mutant of Bartlett; plants and fruits resemble parent except that the fruit colour is dark cranberry red and shoots and leaves have a reddish tinge.

'Moonglow' and 'Magness' are two fire blight resistant varieties evolved in USA, Fles is free from grit cells.

Jorgonelle

It is an European type with superior quality and adapted to South Indian Hills like Kodaikanal (warm winter conditions).

Starcrimson

Trees are medium sized fairly upright and spreading, Fruits medium sized, oblong ovate, pyriform, dark red change to attractive Crimson red in cold storage. Flesh cream white, moderately juicy, aromatic, high in TSS, sweet with excellent eating quality.

Early China

Trees are upright and compact. Fruits round, small, greenish with red blush and very attractive.

Asian types and Hybrids (common pears)

Kieffer

It is well adapted to different climatic conditions and moderately resistant to fire blight. The fruit is brownish, gritty and hard. It is a self unfruitful variety.

Gola

It is found to be suitable for lower altitude. Fruits are large, round and possess excellent keeping quality. Hence it is suitable for long distance transport.

Le Conte

Suitable for lower aremid hills as its chilling require-ment is low. Fruits are round in shape, small in size, yellowish green in colour. But it is a blight susceptible variety.

Patharnakh

This is another low chilling variety. Tolerant to very high temperature and hot winds. It possesses a peculiar quality combination of drought tolerance as well as tolerance to water logged condition. Fruits are round with prominent dots. Fruits have tough skin and hence suited for long distance transport.

Propagation

Pears are commercially propagated by shield or 'T' budding and also by whip and tongue grafting. The root-stocks are raised from the seeds of commercial pear varieties. A number of F1 hybrids of *Pyrus communis* such as Old Home x Farmingdale are multiplied clonally and used as rootstock. Quince (*Cydonia oblonga*) is another related genus which shows very good compatibility with a number of commercial cultivars and it is resistant to wooly aphids, nematodes but susceptible to Oak rooto fungus, fire blight, cold and excess of lime in soil.

Quince clones are easily propagated by semihardwood and softwood cuttings. Some of the commercially used clones of quince are QA, QB, QC. But commercial cultivars of pear like pear Bartlett, Bosc, Eldorado need 'Old Home' as interstock as they don't have compatibility with Quince. The chilled scions of Doyenne du comice grafted at 1.0 m height on 1.0m long shoots of Pant pear –18 stock found to be the best for higher fruit yield and quality.

Preparation of Field and Planting

One year ahead of planting, the field should be prepared by removing stems and roots ofprevious trees and shrubs, leveled giving a gentle slope for drainage of excess water during heavy rains. For a crop on its own rootstock (pear), an initial spacing of 3 Mx2 M is given which is changed to 6 M x 4M after 4-5 years. For pear on Quince, a planting distance of 3.5 M x 1.1 M is enough since quince he has the effect of dwarfing the trees. The pit size should be 1M x 1M x 1M and the pits are filled with a mixture of soil and compost. The planting can be taken up during late fall or early spring. Immediately after planting the basin should be formed and irrigated.

Training and Pruning

Pears are trained in a number of systems like pine shaped, pyramid, spindle, palmette and trellis. Among these, palette system and tatura trellis are found to be commercially superior. In tature trellis, the rows are oriented North-South. Each tree is topped to develop two arms to from 'Y' shape in East – West direction within 500-600 crotch angle. Tensioned wires on steel frames support the arms to a height of 4-5 M and the branches on each arm are trained on each arm are trained on these trellises.

Bearings trees are pruned by combining heading back and thinning out. Pear bears fruit bud on spurs arising on two year old wood and a spur continues to bear for more than six years.

Manures and Manuring

An optimum dose of major nutrients is 600g N, 150 g P and 300g K per tree to get the maximum yield. Normally in pears, the response to P and K can be seen only in soils of low availability of P&K. At higher altitudes where soil pH will be less than7, the 'P' will not be available. Similarly, when the soil pH is more than 7 (alkaline condition) too, the 'P' availability will be less. Under these conditions, application of additional 'P' will increase the yield.

Nitrogen @60g/tree in two splits (2/3rd in January and 1/3rd in May) along with a basal dressing of 40g each of phosphorus and potash was potash was found the best in Bagugosha cultivar of low-chilli pear.

Harvest, Yield and Storage

Fully mature fruits are harvested while still firm and green for distant market. Fort local market, they are left on the trees to get better quality fruits. At an interval of 3-4 days, two orthree pickings are taken up. Fruits should be carefully handled while storage and transit as the bruising is possible by rubbing with one another as well as stalk damage. From a wellmaintained orchard an yield of 30 – 40 tonnes/ ha/year can be expected. The unripe fruits harvested at optimum maturity can be stored for even 5 months at a temperature of –1°C. Ripening can be accomplished by keeping at 15 to 21 and 21 to 25°C and 80-85% RH in 3-6 days depending on the cultivar. Most of the commercial cultivars require this post harvest chilling treatment for proper ripening. When such post harvest chilling treatment for proper ripening. When such post harvest chilling treatment for proper ripening. When such post harvest cold treatment are not available, the fruits can be treated with ethylene, so that they ripen properly and get good quality (both taste and colour).

Chapter 24

Plums & Prunes

Botanical name : *Prunus domestica*
Family : Rosaceae
Origin : China
Type of fruit : Drupe
Edible Portion : Mesocarp
X = 8
2n = 16, 48

SOIL, CLIMATE, PLANTING, VARIETIES, NUTRIENT AND WATER MANAGEMENT, SPECIAL CULTURAL OPERATIONS, PHYSIOLOGICAL DISORDERS, PESTS AND DISEASES, MANAGEMENT PRACTICES

PLUMS Prunus sp.

'Prune' – dried whole without fermentation (high sugar).

Production Yugoslovia – tops in production

India, USA.

Area

- ☆ Sub-tropical plants of Punjab, Haryana, Uttar Pradesh to high hills of Himalaya in Himachal Pradesh, Uttar Pradesh, Jammu and Kashmir.
- ☆ Great scope exists in NE Himachal Pradesh, Jammu and Kashmir, Uttar Pradesh – important states.

Uses

- ☆ Rich in minerals, vitamins, sugars and organic acids, protein, fat and CHO_2's
- ☆ Jam, jelly, marmalades and pies.
- ☆ **Prune :** Plum with high sugar content, dried which fermentation. Dried plums used for chutney preparation Wine and brandy from varieties of high sugar and sorbitol Seed oil -40-50% - cosmetic and medicinal value.

Climate

- ☆ More and wide adaptability of many species and cultivars adapted to different ecology;

Cold winters

Hot summer suitable

High rainfall

Environment

- ☆ Warm winter - Bending branches with holding irrigation
- ☆ Light winter - reduce apical dominance Rain summer pruning Chemicals spray, rootstocks
- ☆ Japanese plum – sensitive to frost (spring flowering) if no spring frost – good.

Rainfall

- ☆ As rainfed crop, 100-125 cm rainfall, well distributed throughout the year.
- ☆ High wind velocity – not suitable
- ☆ Hail prone areas – not suitable.

Soil

- ☆ Avoid water logged, poorly drained, very shallow soils with high salts.
- ☆ Soil depth should be 1.5 m. High alkalinity (5 mg/100 g) + acidic soil à toxic, apply lime.

Origin

- ☆ European plum – Europe
- ☆ Japanese plum - China
- ☆ Cherry plum - Europe, West Asia
- ☆ American plum - N. America
- ☆ Pollinizer should be planted in every 3rd row as 3rd tree.
- ☆ One bee colony/acre – improves fruitset.

Flower Initiation

- ☆ Flower initiation in one season and flowering seen in the next year, flower development greatly affected by age of wood, position of bud on tree, temperature, water, CH_2O, N and others, PGR.
- ☆ Flowering takes ploace on one year old shoot and 2 year old spurs.

Propagation

- ☆ Cuttings – heard, semi-hard wood and soft wood IBA treatment in winter, rooting in 4-6 weeks.
- ☆ Clonal

Rootstock

- ☆ Myrobalan 29°C, Myrobalan – 2261, GF 1246.
- ☆ Peach - Suits to light soil, resistance to nematodes Almounds - Suits to deep soil, but graft incompatibility Apricot - high immunity to nematodes Rootstocks seeds - dormancy problem is seen, so stratification is necessary @ 3-5°C.

Planting

- ☆ Rootstock vigour and soil fertility decides spacing.
- ☆ Square system – suits to all situations.
- ☆ Hexagonal system – fertile and expensive lands.

Training Systems

- ☆ Branching at 1 m height is practiced Training system depends on growth habit
- ☆ Spreading type – open centre Upright type – Central modified leader Open centre – more common in Europe Modified leader – more common in India

Other Systems

- ☆ Hedge rows
- ☆ Pyramid for mechanical farming
- ☆ Palmette

Pruning

- ☆ Intensity differs with varieties
- ☆ Japanese plum – over bearing – heavy pruning
- ☆ European plum – light pruning to renew old spurs.

Fruit Thinning

- ☆ Increase fruit size
- ☆ Increase colour and quality.
- ☆ Reduce breaking of limbs.

Harvest

- ☆ Total economic life span 25-35 years.

Maturity Indices

1. Colour development
2. Flesh firmness
3. Days after full bloom
4. TSS – 12.5°B for European plum
5. TSS/acid ratio – 12 to 15

Post-harvest Management

- ✰ For distance market harvest in the early morning or in Afternoon and leave it for over might cooling.

  CA storage O_2 – 2-3%

  2-3 months

  CO_2 – 2.8%

Chapter 25
Peach & Nectarines

Botanical name : *Prunus persica Batsch*
Family : Rosaceae
Origin : China
Type of fruit : Drupe
Edible Portion : Mesocarp and epicarp
X = 8
2n = 16

SOIL, CLIMATE, PLANTING, VARIETIES, NUTRIENT AND WATER MANAGEMENT, SPECIAL CULTURAL OPERATIONS, PHYSIOLOGICAL DISORDERS, PESTS AND DISEASES, MANAGEMENT PRACTICES

This is a stone fruit in warm temperate climate also grown in sub-tropics, but of inferior quality. It is most popular because of its attractive colour, excellent quality and taste. Grown inwarm temperate zone of Europe, North America, South Africa, Asia and Australia. Nectarines are smooth skin mutants allied to peach. It is non-pubescent peach of smaller size.

Use : Favourite table fruit, highly valued for its taste, nutritive properties and therapeutic uses. Canned, dried, frozen peaches, jam, juice and beverages are popular. Nutrient rich fruit, used as baby food also.

Nutritive Value

- ☆ Good source of sugars, vitamins and minerals.
- ☆ TSS - 8-13°B
- ☆ Total sugars - 8%
- ☆ Ascorbic acid - 6-13 mg
- ☆ Carotenoids - 0.19-0.53% - white flesh cultivars 0.75-0.79% - yellow flesh cultivars.

Peach Kernel Contains

- Fats, proteins, fibre and minerals

  39-55% Fat

  23-30% Proteins

  14.8% Crude fibre

  2.7% Minerals

Glycosides

- Prunacin' – Pulp
- 'Amygdalin' - Seeds

Medicinal Properties

- Peach kernel oil - Food, cosmetics, cattle feed, pharmaceuticals, bio-fertilizer
- Flower & leaves - Pulgative and anthelmintic, urinary stone, kidney function, indigestion

Origin : China

- **India :** Uttar Pradesh, Jammu & Kashmir, Himachal Pradesh, Punjab, Nilgiris
- North eastern staters *viz.*, Arunachal Pradesh, Meghalaya, Manipur. P. behmi a natural hybrid of almond and peach used as rootstock for plum, peach, almonds Nectarinc P. persica var. nucipersica.

Morphological Description

- Fruit is low headed, wide spread tree, drupe from superior ovary and velvety skin.

Climate and Soil

- Limiting factors : Low winter temperature
- Spring frust
- Hail storms
- High humidity
- Chilling hours at 7.2°C more effective
- Temperature < 2-3°C – does not break dormancy.
- High temperature in December – January bloom abnormality and failure of fruit set is noticed.
- High temperature is winter prolongs dormancy and multifies chilling requirement.

Soil

- ☆ Mild to moderate steep hill slopes are ideal
- ☆ Deep valleys not prefused, because sensitive to water logging.
- ☆ Sandy soils with a depth of 7.0 m is ideal.
- ☆ Oxygen supply to the root zone is important, so compact soils with <10-20% pores should be avoided.
- ☆ Ideal pH – 5.8 to 6.8.

Propagation

- ☆ T-budding is practiced.
- ☆ Autumn season à preferred.
- ☆ Leafy succulent soft wood and hardwood cuttings also preferred.

Peach itself is a Successful Rootstock

- ☆ For raising rootstock, seeds from late cultivars are used. If early cv. Is used – germination is poor.
- ☆ Removal of seed coat GA3 treatment – reduce stratification period.
- ☆ Nematode attack is common. But 'Nenaguard and Okinoura' exempted.

Others Rootstock

- ☆ Apricot *P. armeniaca* – Root knot nematode rest.
- ☆ Almond *P. amygdalis* – dwarfing rootstock
- ☆ Western sand cherry – *P. berseyi, P. tomentosa* – dwarfing.
- ☆ Namking cherry – *P salicina* – medicinal to large trees.
- ☆ Micropropagation of both rootstock and scion is possible, axillary shoots and embryo culture are successful.

Planting

- ☆ One year old plants.
- ☆ Spacing 4-6 m in square
- ☆ Spring planting better than autumn planting
- ☆ Trees bare to be white washed to protect it from sun.

Training and Pruning

- ☆ Unpruned trees
- ☆ Tall and dense
- ☆ Weak crotches
- ☆ Surplus scaffold branches
- ☆ Suckers and water sprouts

Various Training Systems

- ✰ Modified leader
- ✰ Open centre
- ✰ V-shaped tatura trellis
- ✰ Pillar
- ✰ High density vase
- ✰ 2=scaffold vase – most efficient for low density (277 to 625 plants/ha)
- ✰ Belgium bench

For HDP

Hedge row - 519-889 plants/ha

Tatura trellis - 666 plants/ha

Medow - 13,333 plants/ha

1st year : Stem cut at 61 cm from ground, 3-4 branches allowed, well spaced and well developed on all sides., All other new growth are removed.

2nd year : 2 well spaced secondary branches on each main branch (Dormant season). 2nd summer : Water sprouts suppressed, Secondary branches encouraged. Forked branches cut to make crotches strong/outside buds are pruned to have speedy shape. In the 3rd dormant pruning diseased, criss-cross branches, water sprouts removed. No severe pruning in early seasons.

II Pruning – bearing trees

Annual pruning

– to maintain open centre

– since it stimulates new wood for production

– It allows penetration of sunlight

– Colour development

– Fungal diseases

Annual heading back

- to maintain low canopy

- 2-3 year old branches to be removed.

- Side branches to be shortened and thinned

- Annual new growth to be maintained @ 41-61 cm length.

Nutrition

- ✰ Balanced nutrition – health and productivity FYM 40 kg/tree.
- ✰ For Indian conditions :

- Nitrogen : 55-65 kg/ha, phosphorus : 55-65 kg/ha, potassium : 110-135 kg/ha bearing orchard.
- NPK : 20:15:15 g/year
- This should be increased annually. Apply fertilizers as band than broadcasting.

Micronutrients

- 0.5% $ZnSO_4$, 0.2% Boric acid and 0.2% CuSO4 corrects respective deficiency.

Irrigation

Soil moisture important particularly at the time of fruit maturity. Deep well drained soils is wetting upto 1.8 m depth. Shallow soils is irrigation at frequent intervals with less water. Stress – less winter hardiness.

Inter Cultivation

- Desirable but confined to 10 cm soil depth by light hoesing or shallow ploughing.
- **Advantages :** Weed competition decreased, good physical condition, soil erosion, soil moisture conserved.
- **Intercropping :** Short duration crops can be grown till peach comes to bearing. Eg: Cowpea, soyabean, turmeric and pineapple.

Green Manuring

- Short duration legumes.
- Incorporate before flowering
- Soil condition will improve and fertility status increased
- Eg: Pear, beans, fenugreek, daincha, sunhemp.
- **Mulching :** Peach normally grown under sod – permanent sod cover, depletion of nitrogen.

Crop Regulation

To regulate heavy flowering and fruiting, thinning can be done to get quality fruits of marketable size.

- Ethrel 50-100 ppm effective in increasing fruit size, fruit weight increased. Pulp – stone ratio increased, twit quality increased
- Spraying : 1000-2500 ppm DNOC prior to full bloom effective.
- Time of thinning : Depends on maturity.
- Early cultivars – blossom thinning.
- Mid & early – at petal fall or fruitset.

Physiological Disorder

1) **Split pit and gumming** – At pit hardening stage splitting at joint of dorsal and ventral side. Exudation of gum and the gum fills the pit cavity and seeds become abortive, fruits become unfit for consumption.

   Causes : prolonged drought followed by sudden rain – temperature and humidity increased – splitting.

2) **Sunscald** – Constant exposure to sun Sunscald on trunk, twigs and branches.

   Control : Painting the exposed area with time and propride shade.

   Harvest : Maturity at harvesting decides the post harvest quality and storage.

Maturity Indices

- ☆ Days after full bloom
- ☆ Fruit size
- ☆ Fruit firmness
- ☆ Pit discolouration
- ☆ Freeness of pit
- ☆ Ground colour
- ☆ Sugar , TSS-arid ratio.
- ☆ Flowering to maturity = 78-127 days Grand colour change and flesh firmness is the best index.

Yield

- ☆ 7-10 tonnes/ha
- ☆ High Density Planting – 78 MT/ha
- ☆ Climacteric fruit
- ☆ Application of growth retardants Alar (SADH) and ethrel
- ☆ Enhance ripening
- ☆ Increase fruit quality
- ☆ Uniform ripening
- ☆ Alar @ 500 ppm at pit hardening, Ethrel – 100, 700 ppm before harvest – effective

Storage

- ☆ Soft textured high moisture is highly perishable, spoil within 2-3 days.
- ☆ Pre-cooling – reduces field heat.
- ☆ Hydrocooling with running H_2O @ 10°C
- ☆ Precooling + 0°C storage with 85-90% RH à 28-36 days storage.

- ☆ CAS – Very effective
- ☆ Packing material of weed, fibre, jute, plastics can be used.
- ☆ Recently corrugated paper board boxes.
- ☆ Plastic film with entrapped air bubbles

Storage Diseases

- ☆ Brown rot, Whiskers rot, grey mould, green mold, blue mould, black mould.

To Control

- ☆ Radiation treatment, hot H_2O dipping - reduces spoilage.

Chapter 26

Classification of Fruits Based on Climatic Requirements, Horticultural and Botanical Classification

Agroclimatic Zones of India

Agro ecological zone is defined as a land unit in terms of major climates, suitable for certain range of crops and cultivars. Fruits are broadly classified on the basis of their temperature requirement such as, temperate, subtropical and tropical. However, some fruits may be grown under more than one climate. For example, mango is grown under both tropical and subtropical climates. Grape and peach can be grown in both temperate and subtropical regions. The agro climatic zones of India are as follows.

- ✰ Zones Temperate Subtropical Tropical
- ✰ North western North eastern
- ✰ Central southern coastal

Characteristics of Different Zones

Temperate zone

Vegetative and flower buds of most of the temperate zone fruits enter into dormancy in late summer or autumn and require a substantial amount of winter cold before they can resume growth in the following spring.

- ✰ These fruit plants are frost - hardy and even tolerant to snowfall and ice,
- ✰ The soil is mainly shallow and acidic in nature, and
- ✰ Rainfall is adequate.

In India, temperate zone climatic environment occurs at sufficiently high elevations.

Subtropical zone

- ✰ Temperature is of intermediate in nature,
- ✰ Chilling temperature, if occurs, is for a brief spell and beneficial for many fruits,
- ✰ The fruits grown have no distinct rest period,
- ✰ Light frost may occur in this zone,
- ✰ Rainfall varies widely from low to high,
- ✰ Soil reaction is generally neutral,
- ✰ Soil fertility status is not satisfactory,
- ✰ The majority areas have sandy loam and alluvial soil, and
- ✰ The fruits have no marked photoperiodic requirement.

Tropical zone

- ✰ Chilling temperature is practically absent,
- ✰ High temperature occurs during most part of the year,
- ✰ Day and night temperature variation is narrow,
- ✰ Photoperiodic requirement in very insignificant,
- ✰ Rainfall varies widely,
- ✰ Soil type varies greatly (clay loam, alluvial, saline, sandy, laterite *etc*.,)

Besides, the arid region of India which occupies nearly 12 percent of the total land area is of immense importance for fruit growing because of the development of highly efficient irrigation management methods. In India, the hot, arid zone covers over 3.2 lakh.Sq.km.in Rajasthan, Gujarat, Andhra Pradesh, Punjab, Haryana, Karnataka and Maharastra. There is about 70,300 sq. km area under the cold arid zone of Ladadh in Jammu & Kashmir.

These Regions are Characterized by Distinct Agro-climatic Features

- ✰ Annual rainfall is low with erratic distribution pattern and largely confined to the period from July – September with 9-12 rain spells out of 12-20 in the whole year. After April, the vapour pressure deficit is more than 24mb, and exceeds 30 mb during may and June causing heavy evapotranspiration. These conditions result in soil moisture and atmospheric water stress to the plants after the rainy season. The mean annual rainfall varies from 100 mm north-western to 450 mm in eastern boundary of the arid zone in Rajasthan. It varies from less than 30mm to 500mm in the arid zone of Gujarat and from 200mm to 450mm in Haryana and Punjab.
- ✰ High solar radiation incidence (450 – 500 cal per cm2/day) and high wind velocity (20km/hour) result in a high potential evapotraspiration (6mm/day) and high mean acidity index (74-78%).

- ✰ The soils being wind blown, have 85% sand and low organic matter (0.1 – 0.45%) with poor water holding capacity (25-28%) and high infiltration rate (9cm/hr).
- ✰ The groundwater resources are meager and at a great depth (40-50mm).
- ✰ Owing to the poor surface and subsurface drainage, the groundwater in a large part of the area (*e.g.* 60% of arid Rajasthan) is highly saline.

Fruit Growing Regions in India

1. **Temperate zone:** Jammu & Kashmir, Himachal Pradesh, part of Uttar Pradesh, Arunachal Pradesh, part of Nagaland, Nilgiris and Pulney hills in Tamil Nadu.
2. **North western subtropical zone:** Rajasthan, Punjab, Haryana, part of UP, part of Mathaya Pradesh.
3. **North Eastern subtropical zone:** Bihar, Assam, Megalalya, Tripura, part of Arunachal Pradesh and part of west Bengal.
4. **Central tropical zone:** Part of Madhya Pradesh, part of Mahsrashtra, Gujarat, part of Orissa, part of West Bengal, part of Andhra Pradesh & part of Karnataka.
5. **Southern tropical zone:** Part of Karnataka, part of AP, part of TN & part of Kerala.
6. **Coastal tropical humid zone:** Coast of Maharastra, Kerala, Andhra Pradesh, TN, Orissa, West Bengal, Tripura and Mozorm, part of Gujarat along sea and the Indian Islands.

The ICAR, New Delhi, has recognized eight agro climatic zones for effective land use planning.

S.No.	Agro climatic Region Status
1.	Humid western
	Himalayan Region
	J&K, HP, Kumaon and Garhwal in Uttarnchall
2.	Humid Bengal – Assam
	Region
	West Bengal & Assam
3.	Humid Eastern
	Himalayan Region
	Bay Islands, Arunachal Pradesh, Nagaland, Manipur, Mizoram, Tripura, Sikkam, Megalaya & Andaman & Nicobar Islands.
4.	Sub-humid Sutlez-Ganga
	Alluvial plains Punjab, Delhi, UP plains & Bihar.

5. Sub-humid to Humid

   Eastern and south Eastern Islands

   Eastern Madhya Pradesh, Orissa and Bihar.

6. Arid western plains Haryana, Rajasthan, Gujarat, Dadra and Nagar Haveli and Daman and Diu

7. Semi-arid Lava plateaus and central Islands.

   Mahastra, Western Central Madhya Pradesh and Goa.

8. Humid to Semi-arid western Ghats

   Karnataka, TN, KErala, Pandichary and Lakshadweep Islands.

During 1985 – 90, the Planning commission accepted 15 broad agroclimatic zones based on physiography and climate for effective planning.

These Zones are

1. Western Himalayan Region.
2. Eastern Himalayan Region.
3. Lower Gangetic plains Region.
4. Middle Gangetic plains Region.
5. Upper Gangetic plains Region.
6. Trans – Gangetic plains Region.
7. Eastern Plateau and Hills Region.
8. Central Plateau and Hills Region,
9. Western plateau and Hills Region,
10. Southern plateau and Hills Region,
11. East coast plains and Hills Region,
12. West coast plains and Ghats,
13. Gujarat plains and Hills Region,
14. Western Dry Region, and
15. The Island Region.

Under National Agricultural Research Project (NARP) each agro climatic zone is divided into sub-sones. The sub-zones totaling 120, are primarily based on rainfall, existing cropping pattern and administrative units, where in each state is divided into 2-12 sub-zones. These delineations are too many for planning process at National level, having similar agro climatic conditions.

The National Be of soil survey & land use planning (NBSS & LUP) Nagpur, delineated the country into 21 agro Ecological regions, using physiographic, soils, bioclimatic types and growing periods. Growing period (GP) was based on water balance starting with period having precipitation 0.5 excess of evaporation (PET) ending with utilization of 100 mm of stored soil moisture once precipitation falls below PET. In this system of delineations, arid zone had GP of 90 days, semi-arid region corresponded to 90- 150 days of GP and the subarid zone correlated with region having GP between 150 and 210 days. The humid and prehumid zones had GP of 210-270 days and more than 270 days/year respectively.

The Climatic Zons of Tamil Nadu

Tamil Nadu forms the southernmost state of Indian Union located between 8° 20 ` to 13 °10 ` north latitude and 76° 15 to 80° 20` east longitude. The man annual rainfalls is 974mm with the highest contribution from north-logged monsooon 47 per cnet of total rainfall), followed by south-west monsson (32 per cent), summer (16 per cent) and winter (5 per cent). Based on the rainfall, altitude and irrigation souce, the state has been marked into seven c;limatic zones (fig.44) whose characterisitcs are furnished below.

1. North-eastern Zone (Chengai-Anna, North Arcot Sotuh Arcot districts)

The mean annual rainfall is 1100mm out of which 566 mm is received during the North-East monsoon period with frequent occurrence of cycloners. The mean maximum temperature is from28.5 to 38.4° C while the mean minimum temperature is from 21 to 28° C.The major soil types are red sandy loam, clayey loam and saline coastal alluvium. The major irrigation sources are hanks and wells. Since the rainfall is uniformly distributed from July to December, two crop sequences of groundnut followed pulses are practised. In the coastal area rain-fed rice is cultivated. The other crops like sugarcance and millets are raised with the help of well.two crop sequences are followed. In this zone a significant practice of transplanting finger milet, sorghunm and tomatopurely under raiunfed condition is followed. Established mango grooves in drylands are common in this zone.

2. North-Western Zone (Salem and Dharmapuri Districts)

The mean annual rainfall is 875 mm and about 42 per cent of rainfall is received during the South – West monsoon period. The mean monthluy maximum temperature ranges from 30 to 37° C and the minimum temperature is from 19 to 25° C. The minimum temperature usually goes below 15 C in northern parts adjoining Karnataka. The elevation rangers from 800 to 1000 m(MSL). The Cultivation is drylands commences from the month opf June and

3. Western Zone (Coimbatore and Periyar Districts)

Themean annual rainfalls is 720 mm with a contribution of 49 per cent from the North-East period. The mean maximum temperature is from 30 to 35° C while the mean minimum temperature ranges from 19 to 24° C.

The predominant soil types in black soils during September/October . With the receipt of early rain, groundnut is sown in red soils. In black soil areas, cotton for early rains and bengalgram for late rains are raised. In the southern oiart if this zone the rainfal is about 550 mm only and more area is devoted to pastures with hardy tress like white babul. With the help of well and cannal irrigation crops, like cotton, finger milklet and sugarcane and raised.

4. Cauvery Delta Zone (Thanjavur and Tiruchirapalli Districts)

The mean annual rainfall varies from 900-1000 mm, out of which more than 50 percent is received through North_east monsoon period. The main source of irrigation is the Cauvery river. The mean maximum temperature is from ranges from 30.0 to 38.5° C and mean minimum temperature is from 21 to 27° C. The major soil typer is alluvial in the old delta areas while red loamy and it has been rightly called as 'rice bowl of Tamil Nadu'. After the rice crop, pulses are raised with residual soil moisture. In places with supplemental irrigation through wells filter points cotton, groundnut and sesamum are raised as summer crops.

5. Southern Zone (Pudukkottai, Madurai, Quaide- e-millath, Passumpon, Kamarajar, Ramanathapuram, Chidambaram and Tirunelveli)

The topography of zone is undulating. This zone lies on the rain shadow are of the western ghats. The mean annual rainfall is 850 mm with a contribution of about 470 mm from North-East monsoon. The mean maximum temperature varies from 28 to 38.5° C, and the minimum temperature ranges from 21 to 27.5° C. The soils of this region fall under major groups, *viz.*, black, red alluvial and lateritic. Saline coastal alluvial soils are also present in the coastal belt. In black soils only one crop, either cotton or sorghum israised. Direct seeded rice is cultivated under rainfed condition onlight soils. On red soils, groundnut crop is raised. Under garden land conditions, pearl millet and chillies from the major crops.

6. High Rainfall Zone (Kanyakumari District)

The mean annual rainfall is 1460 mm received in 64 rainny days, out of which 38 and 36 per cent are respectively received during South-west and North -East monsoon periods. The mean maximum temperature ranges from 28 to 33,5° C and minimum is from 22 to 26.5 ° C. The soils are deep red loam except the crop grown under rainfed condition followed by tapioca. Plantaion crops like tea, pepper, clove, nutmeg, cardamom and coffee also cultivated on the hills.

7. Hilly Zone

This zone c comprises the hilly region of the Nilgris, the Shevroys, the Yalagiri, the Annamalais and the Palani. The rainfall varies from 1000 mm at the foot of the hills to 5000 mm at the peaks. The mean maximum temperature varies from 15 to 24° C and that minimum ranges from 7 to 13°C. The soil is mainly lateritic. The major crops are cole vegetables, potato, tropical and temperate fruit areas. At the foot of th hills minor millets are the raised by hill tribes. At higher altitudes wheat cultivation is common during winter season.

Index

www.ingramcontent.com/pod-product-compliance
Ingram Content Group UK Ltd.
Pitfield, Milton Keynes, MK11 3LW, UK
UKHW020047200726
13845UKWH00044B/34

9 789390 371549